THE DEVELOPMENT OF BIOCHEMISTRY IN CANADA

E. GORDON YOUNG

The Development of Biochemistry in Canada

UNIVERSITY OF TORONTO PRESS
TORONTO AND BUFFALO

© University of Toronto Press 1976
Toronto and Buffalo
Printed in Canada
Reprinted in 2018

E. Gordon Young taught biochemistry at Dalhousie University, Halifax, from 1924 until 1950, when he resigned to become director of the new Atlantic Regional Laboratory of the National Research Council in Halifax. He retired in 1962.

Library of Congress Cataloging in Publication Data
Young, Elrid Gordon.
  The development of biochemistry in Canada.
  Bibliography: p.
  Includes index.
  1. Biological chemistry – Canada – History.
  I. Title. [DNLM: 1. Biochemistry – History –
  Canada. QW11 DC2 Y71d]
  QP511.5.C2Y66   574.1'92'0971   75-44055
  ISBN 0-8020-2229-4
  ISBN 978-1-4875-7285-3 (paper)

Contents

PREFACE vii

INTRODUCTION 3

ORIGINS AND DEVELOPMENT
AT THE UNIVERSITIES 5
McGill 9
Alberta 16
Toronto 20
Manitoba 26
Western Ontario 26
Dalhousie 27
Montreal 29
Laval 31
Queen's 32
Saskatchewan 34
Ottawa 35
British Columbia 36
McMaster 39
Summary and analysis 39
Pathological chemistry 44
  Toronto 44
  Western Ontario 46
  McGill 47

BIOCHEMISTRY IN AGRICULTURAL
COLLEGES 48

BIOCHEMISTRY IN INDUSTRY 51

BIOCHEMISTRY IN GOVERNMENT
LABORATORIES 55
National Research Council 55
Atomic Energy of Canada Ltd. 59
Department of the Environment 60
  Fisheries Research Board 60
  Forest Products Laboratories 63
Department of National Health and
Welfare 64
  Food and Drug Laboratories 64
  Division of Nutrition 67
Department of Agriculture 69
  Agricultural institutes 69
  Grain Research Laboratory 70
  Animal Research Institute 71
  Food Research Institute 72
  Chemistry and Biology Research
    Institute 72
  Plant Research Institute 73

Agricultural Research Institute 73
Other research institutes 74
Provincial research laboratories 74

BIOCHEMISTRY IN SPECIAL
INSTITUTES 76
McGill-Montreal General Hospital
  Research Institute 76
Allan Memorial Institute of
  Psychiatry 77
Montreal Neurological Institute 78
The Banting Institute 78
Banting and Best Department of
  Medical Research 82
Charles H. Best Institute 85
Hospital for Sick Children Research
  Institute, Toronto 85
National Cancer Institute 86

BIOCHEMISTRY IN CANADA
IN PERSPECTIVE 89
Endocrinology 89
Neurochemistry 90

Carbohydrate chemistry 91
Phytochemistry 91

BIOCHEMICAL SOCIETIES
AND JOURNALS 93
Societies 93
Journals 97

DISTINGUISHED EXPATRIATED
BIOCHEMISTS 98

FINANCIAL SUPPORT OF
BIOCHEMICAL RESEARCH 101

TECHNICAL PROGRESS 104

REFERENCES 109

APPENDIX: List of biochemists cited,
with academic qualifications 113

INDEX OF NAMES 121

INDEX OF INSTITUTIONS 128

A section of illustrations follows page 50

Preface

No history of the development of biochemistry in Canada has ever been attempted although a history of chemistry published by Warrington and Nicholls in 1949 and a supplementary volume *Chemical Canada* published by Warrington and Newbold in 1970 contain a few facts about the subject. As one of the elder biochemists in Canada and one who has known personally many of those who have worked in this field since the early days, I have been stimulated to compile a history which lays particular stress on the interests and accomplishments in research within Canadian biochemical and other laboratories during the twentieth century.

The present volume, which has emerged after some ten years of interrupted time since retirement and much correspondence, represents a personal appraisal of these activities. The author regrets if he has omitted to mention investigations of notable merit, and hopes that many of the facts cited will be of interest and value to biochemists practising in Canada now. The past is so quickly forgotten, yet from it is the present and the future. A science is maturing when it has a past worthy of acknowledgment even although it may lie buried in the libraries of the world and in buildings that have now been demolished.

Publication of this history would not have been possible without the financial assistance of the Medical Research Council and the National Research Council of Canada. I am very grateful for their support.

EGY

THE DEVELOPMENT OF BIOCHEMISTRY IN CANADA

Introduction

Biochemistry may be said to have been recognized as an independent discipline with the publication of the first numbers of Maly's *Jahresbericht über die Fortschritte der Thierchemie* in 1873, the *Zeitschrift für physiologische Chemie*, edited by Felix Hoppe-Seyler, in 1877, and Hoppe-Seyler's textbook in 1881. It was thus mainly of German origin, and it is therefore not surprising that many of the earlier North American biochemists took part of their post-graduate training in Germany.

Biochemistry is an interdisciplinary subject, incorporating fragments of knowledge from many earlier disciplines, mainly chemistry and physiology, but also agriculture, biology, and medicine. Because of its fundamental role in biological phenomena biochemistry now invades microbiology, clinical chemistry, immunology, endocrinology, oceanography, nutrition, household science, and some aspects of engineering. It is the scientific basis of much of the pharmaceutical, tobacco, fermentation, and food industries, and of some phases of the chemical industry. Biochemistry has grown faster than any other branch of chemistry in the past fifty years, as shown by the number of publications abstracted in *Chemical Abstracts*.

In 1930 R.H. Chittenden (1), the father of biochemistry in the United States, wrote a book on *The Development of Physiological Chemistry in the United States*. There is no comparable story of the development in Canada except for a very brief and incomplete treatment by A.B. Macallum written for the fiftieth anniversary of the Royal Society of Canada in 1932 (2). As it has been my good fortune to have known personally most of the more prominent Canadian biochemists in the earlier part of the twentieth century, I am here attempting to trace that development from its beginning till the present time.

4 The development of biochemistry in Canada

The growth in the universities will be considered first, and the story will then be taken beyond the university to agricultural colleges, industry, and the special research institutes - federal, provincial, and private - who have contributed so much. Along the way, brief biographies will be provided for more senior biochemists as well as an outline of their main research accomplishments. The volume later touches on the technical developments in the subject in the twentieth century, biochemical societies and journals, and the financial support of research in biochemistry. The list of references includes only those cited in the volume as details of specific investigations are available in the abstracting journals.

Origins and development at the universities

The critical report of Abraham Flexner (3) on the medical schools of North America in 1910 was probably responsible for the birth of most university departments of biochemistry in Canada under pressure of conformity, and for their being linked mainly with faculties of medicine. Table 1 records the year of establishment of separate departments in Canadian universities, but, of course, the teaching of the subject began earlier in most colleges.

The records of McGill University indicate that R.F. Ruttan (q.v.) assisted Sir Wm. Osler in organizing a course in clinical and physiological chemistry in the Department of Chemistry of the Faculty of Medicine, during the summer session of 1883. This was probably the first in Canada and is comparable to the first course in the United States given by R.H. Chittenden at Yale in 1874. Ruttan was a chemist working towards the MD degree at McGill at the time, and probably much of the credit for this venture belongs to him under the stimulus of Osler.

The father of biochemistry in Canada is undoubtedly Archibald Byron Macallum (1858-1934) (4, 5) who was appointed to the first professorship of biochemistry in a separate department in 1908 at the University of Toronto after considerable opposition. Formation of this department is antedated only by that at the University of Liverpool in 1902 and by several at universities in the United States and Germany.

Macallum was a natural leader as is evident from his career. He was born on a farm near Belmont, Ontario. In his childhood at home he heard and spoke only the Gaelic as his father had come, in 1830, from the highlands of Scotland to Canada, where he brought up a family of twelve children. Three of the sons became prominent doctors of medicine. 'A.B.' attended a rural school and then the

6 The development of biochemistry in Canada

TABLE 1
Dates of establishment of departments of biochemistry and of pathological chemistry, and the first heads of the departments

University	Year	Head of department
BIOCHEMISTRY		
Alberta	1920	J.B. Collip
British Columbia	1950	M. Darrach
Dalhousie	1924	E.G. Young
Laval	1940 (Medicine)	J.R. Gingras
	1943 (Science)	E. Bois
Manitoba	1923	A.T. Cameron
McGill	1922	A.B. Macallum
McMaster	1967	R.H. Hall
Memorial	1967	L.A.W. Feltham
Montreal	1951 (Medicine)	G.H. Baril
Ottawa	1946	V. Vlassopoulos
Queen's	1937	R.G. Sinclair
Saskatchewan	1946	H.B. Collier
Sherbrooke	1968	R.H. Despointes
Toronto	1908	A.B. Macallum
Victoria	1968	A.J. Wood
Western Ontario	1921	E.G. Young
PATHOLOGICAL CHEMISTRY		
Toronto	1910	J.B. Leathes
Western Ontario	1951	E.M. Watson

high school in London, Ontario, where he attained a first-class teaching certificate before he was seventeen. He then taught school for several years before achieving his baccalaureate together with the medal in natural science in 1880 at the University of Toronto.

For three years he taught in the high school at Cornwall in eastern Ontario. There he met his wife-to-be and also made a lifelong friend of James Whitney, a young lawyer, who later became premier of Ontario and through whom Macallum was able to exert an influence on education in that province.

In 1883 Macallum was appointed lecturer in biology at the University of Toronto to teach physiology and biochemistry under Ramsay Wright, who was head of the Department of Biology at that time. Later he took a PH D at Johns Hopkins University under H. Newell Martin in 1888 and the MB at Toronto in 1890. He was then appointed to the new professorship of physiology at the University of Toronto and later transferred to the professorship of biochemistry when established in 1908. In 1917 Macallum became the first chairman of the National Research Council of Canada and in 1920 he accepted a professorship at McGill. A separate department was established for him in 1922 on his return from a trip to China where he had gone to lecture and assist in the organization of the Peking Medical School under the auspices of the Rockefeller Foundation. 'A.B.,' as he was often called, remained at McGill until retirement in 1928, when he went to live with his son Bruce in London, Ontario.

Although the volume of Macallum's research work is comparatively small, he contributed to the localization of the elements calcium, potassium, and iron in plant and animal tissues by microchemical tests. He also made a comparison of absolute and relative concentrations of the inorganic elements in sea-water and in the body fluids of many animals which supported the concept of the origin of land animals from the sea. This work was recognized by his election to Fellowship in the Royal Society of London in 1906, an unusual honour for a Canadian at that time. He played a prominent part in the founding of the American Society of Biological Chemists in 1906 and served as its president in 1911-13. He was president of the Royal Society of Canada in 1916-17 and Flavelle medallist in 1930.

He was a man of striking appearance and forceful personality, slender but over six feet in height, with a beard, moustache, and leonine head. He was a hard worker and had a keen intellect. He was outspoken in discussion, which made him both feared and revered. I recall an incident when the American Society of Biological Chemists was holding its annual meeting in Toronto in December 1922 and A.P. Mathews of Chicago contributed a speculative paper entitled 'Concerning the Nature of Cohesion.' At its conclusion Macallum rose at the back of the room and emphatically remarked: 'This paper should have no place in the programme of our society.' Macallum had a liking for metaphysical discussion and possessed a retentive memory. In his youth he had memorized many passages of prose and poetry so that his contributions to any discussion were usually very colourful. It has been said that on the first occasion of registering at the university he was asked for his name and initials. He replied, 'A for Archibald, Macallum.' On being pressed for a second initial he was rather taken aback but invented B for Byron without the benefit of baptism. He attracted few postgraduate students but he exerted great influence in educational matters in Canada, especially in medicine.

In the faculty at Toronto he became closely associated with two former students of Ramsay Wright, J. McMurrich in anatomy and J.J. MacKenzie in pathology, as well as with A. McPhedran in medicine and I. Cameron in surgery. This group with Macallum as leader transformed the outlook of the faculty and sought to establish a sound foundation based on the biological sciences but against hot opposition from the clinically minded members.

8 The development of biochemistry in Canada

In 1909 Alexander Thomas Cameron (1882-1947) (6) commenced teaching biochemistry in the Department of Physiology at the University of Manitoba where Swale Vincent was the professor. A sub-department of biochemistry was set up in 1920 and an independent department in 1923 with Cameron in charge. He was born in London, England, of Scottish parentage, and was educated at the University of Edinburgh. As an 1851 Exhibitioner he worked for two years with Sir Wm. Ramsay at University College, London, and one year with Fritz Haber at the Technical High School in Karlsruhe. In 1909 he came to Canada as lecturer in physiology at Manitoba. He remained in Winnipeg until his death in 1947, except for a summer of research with A. Kossel at the University of Heidelberg and three years as captain, RAMC, during World War I when he served as chemist officer for water purification with the British Expeditionary Force in France.

Cameron published numerous papers on the biochemistry of iodine and calcium. 'A.T.,' as he was known to friends and students, was a meticulous analyst. During the summer of 1921 he studied the concentration of chloride, bromide, and iodide in two samples of sea-water from the Strait of Georgia, between Vancouver Island and the mainland of British Columbia. With Irene Mounce he analysed several hundred samples of sea-water in that area to determine the physical and chemical factors that influence the distribution of marine flora and fauna in these waters. Cameron was thus one of the earliest chemists to do oceanographic studies on Canada's west coast. These were followed by a long series of papers on the effect of thyroid and thyroxin feeding on rates of growth and hypertrophy of the organ in the rat. Another series dealt with the biochemistry of calcium in hyperparathyroidism and tetany. Cameron's last interest led to the writing of a short monograph, *The Taste Sense and the Relative Sweetness of Sugars and Other Sweet Substances*, which embodied the results of three years of work in this field. His *Textbook of Biochemistry* (1928), the first Canadian one to be written, appeared in six editions as well as in Chinese and Spanish translations. Cameron also wrote *Recent Advances in Endocrinology* (1933) and *Biochemistry of Medicine* (with C.R. Gilmore, 1933). Cameron served as chairman of the Fisheries Research Board of Canada from 1933 to 1946 and devoted much time to it. One of his lifelong interests was the Scientific Club of Winnipeg, of which he was an early member and the secretary for twelve years, and to whose *Proceedings* he contributed forty-six papers.

He had a strikingly original personality and a caustic wit, which was rather repellent to casual acquaintances; his manner was more cordial to established friends. He exhibited the characteristic reserve of the Englishman. Cameron was a heavy smoker. Whether reading or writing, weighing or titrating, a cigarette was usually between his lips. Intent on his work, he allowed the ash to grow longer

Origins and development at the universities 9

and longer. Often it fell onto his clothes or papers. Observers feared that it might fall into apparatus, but at the bench he seemed to have a fine sense for anticipating the critical moment. Cameron was elected a Fellow of the Royal Society of Canada in 1920 and was president of Section V in 1929-30. He was awarded the CMG in 1946.

MCGILL

Robert Fulford Ruttan (1856-1930) (7) was first appointed in 1887 an assistant to Professor G.P. Girdwood in the Department of Chemistry of the Faculty of Medicine and later became officially professor of organic and biological chemistry (1902-22) within that department. L. Bauman from 1909 to 1911 and then V.J. Harding from 1911 to 1920 were his assistants on the biochemical side. Ruttan may be considered a pioneer at McGill, not only for his role in organizing the course in clinical and physiological chemistry, but also for the stimulus which he gave to research and the teaching of post-graduate students. He did very little work in the laboratory himself and only published a few papers on the composition of adipocere and bog butter.

Ruttan was born at Newburgh, Ontario, in 1856. His remote forebears were Huguenots and his grandfather a United Empire Loyalist. His father graduated in medicine from McGill in 1852 and his mother, Caroline Smith, came from Montreal. Ruttan matriculated into University College, Toronto, and at the age of twenty-five graduated wih a BA degree and the gold medal in the natural sciences. Delay in graduation was due to preoccupation with sports and scholastic failures which caused his father to withdraw his financial support. In consequence Robert taught school for two or three years before completing his baccalaureate in 1881. He then entered McGill and graduated with the MD, CM in 1884. Two years of research with A.W. von Hofmann in Berlin in organic chemistry were followed by appointment to the staff at McGill in 1887, where he stayed for forty years until retirement in 1927. From 1891 until 1902 he served as registrar of the Faculty of Medicine. In 1912 he succeeded Wallace Walker as Macdonald Professor of Chemistry and head of the amalgamated departments. He was elected a Fellow of the Royal Society of Canada in 1895 and served as president of the society in 1919. He was the first Canadian to become president of the Society of Chemical Industry in 1921. He died in 1930 in his seventy-fourth year.

Ruttan was an exceptionally good administrator and an enthusiastic teacher of elementary classes. He was affectionately known as 'Bobby' by his students. His was a dominating figure and personality on the campus. He was a bachelor, a clubman, and a sportsman. He was a charter member of the University Club and also a member of the St James and Mount Royal clubs of Montreal, the Rideau Club of Ottawa, and the Chemists' Club of New York. In sports he was a long-distance runner, a yachtsman, and a cricketer in his early days and an enthusiastic golfer in later life. Over six feet in height, he walked erect and was always dressed in tailored lounge suits. His attention to the girls in his classes was a constant source of amusement to the male members. He must be recognized as a pioneer in biochemistry in Canada along with his friend A.B. Macallum.

Victor John Harding (1885-1934) (8-10) was born in England near Bury, Lancashire, and educated as an organic chemist under W.H. Perkin Jr at Owen's College, Manchester. In 1911 he went to McGill as lecturer in chemistry and in 1917 was made associate professor of physiological chemistry. In 1920 he left McGill to become professor of pathological chemistry at Toronto where he remained until his death in 1934. Harding's particular field of research was the metabolism of carbohydrates and ketogenesis in pregnancy and its toxaemias. His early papers on the nausea of pregnancy introduced the widespread clinical use of glucose therapy, and his studies in dehydration entered an unexplored area in biochemical knowledge. Later he made important contributions to the problems of the toxaemias of pregnancy and the effects of hypertonic saline published in 1930 and 1932. His migration from research in pure organic chemistry to the chemical pathology of obstetrics is most remarkable in one entirely lacking in didactic clinical or biological training. Yet his investigations of the metabolism of pregnancy pioneered biochemical interest in the subject and had a very definite influence in the rationale of therapy.

Harding possessed a subtle sense of humour, Gilbertian and kindly. Essentially shy, modest, and sincere, there was in him a deep compassion for humanity and sympathy for its weaknesses. Apart from the laboratory he mingled diffidence with disarming intimacy. Politics and religion were mentioned rarely. He was an excellent lecturer and followed the practice, unusual at the time, of documenting his statements. He was elected a Fellow of the Royal Society of Canada in 1923. Harding attracted and trained numerous post-graduate students who later became prominent biochemists either in Canada or the United States, including R.M. Archibald, A.S. Cook, T.G.N. Drake, B.A. Eagles, O.H. Gaebler, G.A. Grant, S.H. Jackson, T.F. Nicholson, and E.G. Young.

Harding's assistant at McGill, George E. Simpson (1889-1927), an American who had graduated from Yale University under F.P. Underhill and done research with V.C. Myers at the New York Post-graduate Medical School, carried on as assistant professor from 1920 until the arrival of A.B. Macallum in 1922. He left for Pennsylvania in 1924 and was replaced by Sydney Bliss (1892-1960), an American from Harvard, who stayed from 1925 until 1929.

Subsequent to the periods of Macallum (1922-28) (q.v.) and of J.B. Collip (1928-41) (q.v.) at McGill, David Landsborough Thomson (1901-64) (11) directed the department between 1941 and 1959. He had been closely associated with Collip in endocrinological research since 1928. He was born in Aberdeen, Scotland, the son of Sir J. Arthur Thomson, Regius Professor of Natural History at the university. He was educated in Scotland and England where he worked under F.G. Hopkins at Cambridge. He migrated to McGill immediately after qualifying for his doctorate in 1928 and remained there until his untimely death in 1964. After regular promotions he became professor in 1936 and head of the department in 1941. He also served as dean of the Faculty of Graduate Studies (1942-61) and as vice-principal (1955-61).

David Thomson was a very distinguished man in several characteristics. He became one of the 'features' of the McGill of his time, known to thousands of students with pleasure and gratitude. He had a colossal memory and was a superb speaker, writer, and raconteur. Included in his list of publications is a little book entitled *The Life of the Cell* and another non-scientific one, *Murder in the Laboratory*, under the nom-de-plume of T.L. Davidson. He read, absorbed, and retained the pertinent current biochemical literature. The high quality of his lectures attracted many students to biochemistry. As dean of Graduate Studies he took the chair at all oral examinations for the doctorate and astonished candidates and colleagues by his pertinent questions regardless of subject.

Latterly Thomson became so involved in administration of the university and associated responsibilities that his departmental colleagues had to do the practical teaching with deficient equipment in inadequate laboratories. He was never very active himself in the laboratory but collaborated with Collip and his associates in the preparation of scientific articles. He left to others the direction of research work in the department, mainly O.F. Denstedt from 1937 to 1967, R.D.H. Heard from 1942 to 1957 on steroidal hormones, and J.H. Quastel (q.v.) from 1946 to 1966 on many aspects of cellular metabolism. Thomson served on several national councils.

In appearance Thomson was lean, elegant, and attractive. He was a good conversationalist, with an infinite collection of anecdotes. His extracurricular specialties included a love of music and of mountaineering, a purist attitude in the mixing of martinis, and a considerable skill at billiards. In November of 1961 he fell on ice and struck his head, which inflicted an injury that was complicated by a rare blood dyscrasia. Despite heroic efforts his speech was lost and cerebral degeneration ended in death in October 1964.

Orville Frederick Denstedt (1899-1975) was born in Blyth, Ontario, earned his BSC at the University of Manitoba in 1929, and was an assistant in the Pacific

Fisheries Experimental Station, Prince Rupert, British Columbia, from 1929 to 1932. The use of drying fish oils in protective coatings, especially paints and varnishes, and the chemical composition of fish oils occupied his attention at this period. He returned to the east, received his PH D from McGill in biochemistry (1937) under Collip, and became a lecturer in the department. He was appointed assistant professor in 1942, associate professor in 1946, full professor in 1961, and Gilman Cheney professor from 1965 to 1967 when he retired. He served as president of the Canadian Physiological Society (1955) and of the Canadian Biochemical Society (1959), and was elected a Fellow of the Royal Society of Canada in 1964. With his students Denstedt investigated blood preservation, enzymology of red blood cells, capillary fragility, the biochemistry of vascular disease, liver function, the chemistry and physiology of the hormones and the lipids of the anterior and posterior pituitary, melanophore hormone, some aspects of iron, zinc, and silicon metabolism, and the biochemical action of insecticides. He is best known for his work on blood and the metabolism of erythrocytes.

Robert Donald Hoskin Heard (1908-57) directed studies on the chemistry and metabolism of gonadotropic hormones during a period when the relationships of these steroidal compounds were far from clear. He was born in St Thomas, Ontario, and received his BA in 1929 and MA in 1930 from the University of Toronto. He was an 1851 Exhibition Scholar in 1930-33 and received his PHD from the University of Manchester in 1932 under H.S. Raper. He then spent a year in Oxford. He was a Banting Research Foundation grantee in Toronto (1933-35) and a research assistant in the Connaught Laboratories (1935-37) before becoming assistant professor of biochemistry at Dalhousie in 1937. In 1942 he went to McGill as assistant professor and died prematurely in 1957.

Kenneth Allan Caldwell Elliott was appointed to succeed Thomson as head of the department in 1959. He was born in Kimberley, South Africa, in 1903. In 1924 he received an MSC from Rhodes University and then worked for a year as chemist with the Rhodesia Broken Hill Development Company. Two years as chemist in a dynamite factory in South Africa made possible a period in Cambridge, where he earned a PHD in biochemistry in 1930. Elliott was research chemist in the Biochemical Research Foundation of the Franklin Institute (1933-39) and in charge of chemical research in the Laboratory Institute of the Pennsylvania Hospital (1939-44) before coming to McGill in 1944 as assistant professor of biochemistry and neurochemist at the Montreal Neurological Institute. Here he did notable research on the assay and role of γ-aminobutyric acid and other substances with neuro-inhibitory properties and on respiration and glycolysis in brain slices. He retired in 1969 and returned to Africa to the University of Nigeria in Enugu. During his period in Canada (1944-69), Elliott served as editor of the *Canadian Journal of Biochemistry* for two years. He was elected a Fellow of the Royal Society of Canada in 1963.

Angus Frederick Graham succeeded Elliott as head of the department in 1970. A Torontonian by birth and education, he took a PHD in biochemistry and a DSC in microbiology at Edinburgh University in 1942 and 1952 respectively. From 1940 to 1947 he was a lecturer in biochemistry there. On his return to Canada he was a research associate at the Connaught Laboratories (1947-58) and then at the Wistar Institute in Philadelphia (1958-70) with professorial rank at the University of Pennsylvania. His scientific interest has been the replication of mammalian viruses.

Esau Abras Hosein has been associated with the Department of Biochemistry since 1952 when he was appointed lecturer after obtaining his PHD under Denstedt. He has made many original contributions in the field of neurochemistry, including investigation of the metabolism of simple bases – anserine, carnosine, carnetine, betaine, and their derivatives in nervous tissue – and their activity in the convulsive state in experimental animals. He has accomplished the isolation of acetyl-l-carnityl CoA, γ-butyrobetaine, crotonbetaine, and carnitine from brain tissue in epilepsy.

David Rubinstein was appointed a research assistant at McGill in 1950. Born in 1929 in Montreal, he was educated at McGill where he took his PHD in 1953 and his MD in 1957. After one year at the Michael Reese Hospital in Chicago he became associate professor of biochemistry in 1959 and progressed to full professorship in 1971. He has also been associated with research in the Department of Experimental Medicine at the Royal Victoria Hospital since 1968. For several years he studied the toxicity of carbon tetrachloride and other chlorinated compounds, but his primary investigations have been in the field of the metabolism of erythrocytes, especially the regeneration of ATP in preserved blood. He has also contributed to means of separation and the biosynthesis of lipoproteins in intermediary metabolism.

Murray Judson Fraser was appointed to the staff in the Department of Biochemistry in 1967. He was born in Yarmouth, Nova Scotia, in 1930, and received his early university education at Dalhousie. As an 1851 Exhibition Scholar he went to Cambridge University and qualified for the PHD degree in colloid science in 1957. After two years as assistant in the National Institute for Research in Dairying at Reading he returned to Canada to the McGill-Montreal General Hospital Research Institute. Fraser then joined the staff of the Manitoba Cancer Research Unit with a simultaneous appointment in the Department of Biochemistry of the University of Manitoba (1959-64). In 1964 he moved to the Ontario Cancer Institute in the section of medical biophysics and in 1967 to McGill as associate professor. His research interest has been in the field of nucleic acids: the isolation of repetitive sequences from DNA and the action of nucleases specific for single-stranded acids. For five years (1967-72) he was editor of the *Canadian Journal of Biochemistry*.

The first course in physiological chemistry at McGill was given in the Medical Building in 1883. Even after the fire and the construction of a new medical building on Pine Avenue the laboratories in what became known as the Old Medical Building continued to be used. This was still true after its reconstruction when it was called the Biological Building. Only in 1965 was the department moved into the new fifteen-storey McIntyre Medical Sciences Building on Pine Avenue.

In 1924 the Rockefeller Foundation donated $500,000 to establish the University Clinic at the Royal Victoria Hospital within the Department of Medicine. It was designed to foster original research under the direction of Jonathan C. Meakins who had just returned from the Barcroft expedition to the Andes which had established the direct relationship of concentration of haemoglobin to altitude. He was assisted by C.N.H. Long (1901-74), an Englishman, who later became Stirling Professor of Biochemistry at Yale University.

When Meakins left McGill for Edinburgh in 1947 he was succeeded by John Symonds Lyon Browne, a native of London, England, born in 1904, but who had received his university education at McGill (BS, 1926; MD, 1929; PHD, 1932). He took a doctorate in biochemistry under Collip, working on the isolation of oestriol from human placenta, which focused his lifelong interest in endocrinology. After one year of research with A.F.J. Butenandt at the University of Göttingen he was appointed a lecturer in the Department of Medicine in 1933 and became director of the University Clinic in 1947. In 1948 he was also appointed to full professorship in a new Department of Experimental Medicine designed to accommodate post-graduate students working towards the MSC and PHD degrees and recognized by the Faculty of Graduate Studies. Because of failing vision Browne was later made head of a Department of Investigative Medicine created for him in 1955 so that he could continue his work with less responsibility. He retired in 1969 and the department was disbanded. For forty years Browne was assisted by Eleanor Venning in the investigation of many aspects of biochemical and clinical endocrinology, especially of the steroids isolated by Collip and his associates. During World War II a group led by Browne studied protein metabolism in cases of shock as in aviators who had suffered burns. Other fields of investigation included salt balance in Addison's disease, excretion of sex hormones in clinical cases, and the metabolism of histamine. He was the organizing secretary of the Canadian Physiological Society in 1935 and was elected a Fellow of the Royal Society of Canada in 1939.

Eleanor Venning was closely associated with Browne until retirement in 1968, initially as a research fellow in the Department of Experimental Medicine, with subsequent promotions to full professorship in 1960. She was elected a Fellow of the Royal Society of Canada in 1955.

An important invention made by Kenneth A. Evelyn when he was a research fellow in the department deserves mention. It was a photoelectric colorimeter with filters, one of the first of such instruments to be constructed. It was manufactured commercially and used extensively for many years. The details of the apparatus were published in the *Journal of Biological Chemistry* in 1936-37. Evelyn was a Jamaican by birth who served for five years in the RCAF and afterwards went into medical research at McGill and then in Vancouver.

The work on steroids expanded to such a degree that when Browne retired as director of the University Clinic a separate steroid laboratory in the Royal Victoria Hospital was established.

The director since 1965, Samuel Solomon, has made significant contributions to the pathway of formation of androgens and oestrogens. He was the first to demonstrate the conversion of progesterone to 17 α-hydroxyprogesterone and Δ^4-androstenedione in ovarian tissue and in human foetal adrenals and the first to identify 20/α-hydroxycholesterol as an intermediate in the biosynthesis of pregnenolone. He has an international reputation among steroid biochemists. He was born in Brest-Litovsk in 1925 but graduated from McGill and obtained his PHD in 1953 under R.D.H. Heard. He is also full professor in the Departments of Biochemistry and Experimental Medicine at McGill. Solomon was elected a Fellow of the Royal Society of Canada in 1974.

In 1953 Alec Sehon was appointed an assistant professor in the Department of Experimental Medicine. He was born in Rumania in 1924 but educated at the University of Manchester, where he took his PHD in physical chemistry in 1951. In a sharp change of scientific interest, Sehon began research in the field of immunochemistry in collaboration with Bram Rose. He installed modern equipment and applied the techniques of electrophoresis and ultracentrifugation to the separation and purification of antigens and antibodies. His early work dealt with the antigens in ragweed pollen and the antibody found in the γ-globulins of serum of hypersensitive individuals. Sehon has published many papers in the field, with many collaborators. He has determined the rate constants for the reaction of an antibody with its antigen. He has devised methods of purifying antibodies by the use of corresponding water-insoluble antigens, e.g., bovine serum albumin coupled to diazotized polystyrene, and used this as a means to determine the location of the antibody in the globulin fractions of immune sera. He found it mainly, but not exclusively, in the γ-globulin in very low concentration. Sehon has applied the same technique to myoglobin in an effort to characterize its antibody. In 1956 he was transferred to the Department of Chemistry. In 1969 Sehon became head of a new Department of Immunology at the University of Manitoba and was elected a Fellow of the Royal Society of Canada in the same year.

Mention should be made of Harold Hibbert (1877-1945) (12), who was E.B. Eddy Professor of Industrial and Cellulose Chemistry at McGill from 1925 to 1945. He was born in Manchester, England, and educated at Victoria University, where he did post-graduate research under W.H. Perkins Jr. After qualifying for the PHD at Leipzig under A. Hantzsch, he held several positions both in England and the United States before his appointment to McGill and made notable contributions in organic chemistry. His biochemical research work concerned the bacterial syntheses of polysaccharides, especially cellulose, by *Acetobacter xylinum*, and the structure of lignin. H.N. Brocklesby, A.C. Neish, N.M. Carter, and H.L.A. Tarr were some of his post-graduate students. All four of these men later attained senior administrative posts as directors of biochemical research: Brocklesby, Carter, and Tarr, by strange coincidence, followed one another as directors of the Pacific Experimental Station of the Fisheries Research Board of Canada, and Neish became Director of the Atlantic Regional Laboratory of the National Research Council.

ALBERTA

In 1915 James Bertram Collip (1892-1965) (13-16) was appointed to the staff of the University of Alberta in the Department of Physiology to teach biochemistry. Thus began the professional career of one who may fairly be called our most eminent Canadian biochemist. Collip was born in Belleville, Ontario, on 20 November 1892. At the age of fifteen he entered Trinity College, University of Toronto, registering in the new honours course in physiology and biochemistry. He obtained the BA in 1912, and then took his PHD in 1916 as a student of A.B. Macallum, whom he held in admiration and respect throughout his lifetime. The subject of his thesis was the formation of hydrochloric acid in the gastric cells of the vertebrate stomach. He married Ray Vivian Ralph, a fellow student at Trinity College, before going to Edmonton and his first teaching appointment as lecturer in biochemistry at the University of Alberta. He taught at Alberta from 1915 to 1921, when he was awarded a Rockefeller Travelling Fellowship, which took him back to Toronto to work with J.J.R. Macleod. In the autumn of 1921 F.G. Banting, through Macleod, invited him to join the team engaged on the investigation of insulin, then as a crude extract of pancreatic tissue. Collip quickly increased the amount of ethanol introduced into the extraction procedure from 65 to 80 per cent and thus with other modifications obtained insulin sufficiently pure for use in human patients. The process was patented in the names of Banting, Best, and Collip and turned over to the Governors of the University. Collip remained in Toronto for only a year, as he was increasingly disturbed by the clash of personalities which developed within the team. He, however, had become very interested in the possible development of the field of endocrinology and had demonstrated the syndrome of hypoglycaemia in rabbits and the control of convulsions by intravenous administration of glucose. Prior to his return to Edmonton in 1922, a Department of Biochemistry was established for him with one assistant.

In 1928 he succeeded Macallum at McGill and served as professor until 1941. Then he became Gilman Cheney Professor of Endocrinology and director of a new Research Institute of Endocrinology, where many practical problems concerned with World War II were investigated, such as traumatic shock, motion sickness, blood preservation, and pulmonary irritants. Collip's students and associates were many, including J.S.L. Browne, Evelyn M. Anderson, L.I. Pugsley, C. Gwendolyn Toby, M.K. McPhail, L.W. Billingsley, C.M. Harlow, O.F. Denstedt, A.H. Neufeld, R.L. Noble, and H. Selye. From 1938 he served on the Associate Committee on Medical Research of the National Research Council and as its chairman after the death of Banting with whom he spent the last evening prior to the fatal crash in Newfoundland in 1941. He resigned from McGill in 1947 to become dean of medicine at the University of Western Ontario (1947-61) and director of the Collip Medical Research Laboratory. The latter position he held until his death in 1965, but here he left the direction of research to members of his staff: R.L. Noble, J.C. Paterson, R.W. Begg, J.A.F. Stevenson, C. Engel, K.K. Carroll, and B. Taylor. Up to 1965 some 125 papers had been published from this laboratory. He served as editor of the *Canadian Journal of Medical Sciences* and its successor, the *Canadian Journal of Biochemistry and Physiology*, from 1944 to 1956.

Collip was an indefatigable investigator and one who hated the formal lecture. His major interest was the biochemistry of animal hormones. At Alberta he prepared the first biologically active agent from the parathyroids which he named 'parathormone.' At Toronto he devised the ethanolic method for the purification of insulin. At McGill he worked on the hormones in the anterior pituitary gland, in the placenta, and in the urine of pregnancy. This resulted in the oral oestrogenic drug known as 'Emmenin' and the establishment of the adrenocorticotropic hormone (ACTH) as a separate pituitary hormone.

Collip built a school of biochemical endocrinology at McGill which attracted many post-graduate students who were welded into an effective team of experimenters. During this period some 200 papers were published from his department. He was elected a Fellow of the Royal Society of Canada in 1925 and of London in 1933. During his lifetime Collip received many honours; he was Flavelle medallist in 1936 and served as president of the Royal Society of Canada in 1942-43.

In his obituary of Collip (13), Rossiter draws a picture of a modest, shy, emotional, very energetic man who loved his fellow man: his students, colleagues, and friends. He took pleasure in entertaining them and enjoyed listening to the conversation, contributing only occasionally, but sometimes emphatically and incisively. Bert Collip will be remembered with great affection as an unassuming, warmhearted, self-effacing, and kindly man.

A tribute should be paid here to Mrs Collip, who proved to be an ideal wife for Bert, an understanding woman of quiet charm and great perception. The last episode in his life is typical of the man. He and his wife motored from London to Edmonton, then went to Vancouver by train to attend the annual meeting of the Royal Society. On the return journey Collip drove 1250 miles from Edmonton to Port Arthur in two days to catch the SS *Assinaboia*, which took them to Port McNicholl. Collip then drove the remaining 200 miles to London. Two days later he suffered a stroke and died within a few hours on 19 June 1965, at the age of seventy-two.

When Collip left Edmonton for McGill in 1928, George Hunter succeeded him in 1929 as head of the Department of Biochemistry at the University of Alberta. He was a Scotsman, educated at Glasgow University, who had emigrated to Canada in 1922 as lecturer in pathological chemistry at the University of Toronto under V.J. Harding. He was a good chemist who did meticulous research on the diazo reaction in urine and blood, on certain aspects of the chemistry of imidazoles, and most notably on natural non-protein compounds that contain sulphur. He deserves credit for the discovery of ergothioneine in erythrocytes which he showed to be a source of error in the estimation of uric acid in blood. George Hunter was elected a Fellow of the Royal Society of Canada in 1933, but later resigned in protest at the treatment which he received in Canada.

Hunter resigned his professorship at Alberta in 1949 through a difference of opinion with the administration and some complaints from medical students of the inappropriate nature of some of the lectures in biochemistry. He returned to England in 1950 and has subsequently served as biochemist in several hospitals. He was succeeded by H.B. Collier (q.v.) who transferred from Saskatchewan and who held the professorship until 1961, when he moved to the Department of Pathology to teach clinical biochemistry. Over many years Collier's research interests have been the biochemistry of erythrocytes and the metabolisms of phenothiazine. He retired in 1971.

John Sparby Colter succeeded Collier as head of the Department of Biochemistry in 1961, and it was expanded notably over the following decade in staff, scope of research, and post-graduate instruction. Colter was born in Banff, Alberta, in 1922. He attended the University of Alberta and McGill University, where he qualified for the PHD degree in 1951, and then did research work in the Lederle Laboratories (1951-57) and at the Wistar Institute in Philadelphia (1957-61) before his appointment in Edmonton. His research has been devoted mainly to the biochemistry of viral infections and malignancy. Two other members of the department hold full professorships.

Neil Bernard Madsen was born in Grande Prairie, Alberta, in 1928 and obtained his BSC and MSC at the University of Alberta. He took his PHD in biochemistry in 1955 at Washington University under C.F. Cori, then spent a year at Oxford (1956-57). From 1957 to 1962 he was a research officer in the Microbiological Institute of the Department of Agriculture in Ottawa before his appointment to the staff of his Alma Mater. He has published papers on phosphates and phosphorylases: their activation, inhibition, and relation to glycogen metabolism and the Krebs cycle in *Xanthomonas phaseoli.*

Cyril Max Kay was born in Calgary in 1931. After graduation with a BSC from McGill in 1952, fellowships of the Life Insurance Medical Fund enabled him to take a PHD at Harvard in 1956 under J.T. Edsall and to spend a year at Cambridge University with K. Bailey. In 1958 he was appointed to the staff of the Department of Biochemistry at the University of Alberta. His field of research has been almost exclusively the physical chemistry of proteins, especially those of muscle tissue: tropomyosin, myosin A and G-actin, but also fetuin, trypsinogen, and the fibrogen-fibrin relationship. He has also examined the circular dichroism of ribonucleates.

Lawrence Bruce Smillie was appointed an assistant professor of biochemistry in 1955 after graduation with the PHD from the University of Toronto under G.C. Butler. He was born in Galt, Ontario, and had taken his baccalaureate at McMaster University in 1950. On leave he spent a year at the Laboratory for Molecular Biology in Cambridge (1963-64) and another at the University of Birmingham (1971-72). Smillie was promoted to full professorship in 1967. His published work has been on the structure and function of enzymes, notably chymotrypsinogen A and B, and the amino acid sequences at the active centre of some serine enzymes. Earlier, with colleagues, he examined the properties of calf thymus histone.

Jules Tuba, a Hungarian by birth but educated in Canada at the universities of Saskatchewan and Toronto, was appointed as lecturer in 1942 and professor in 1952. With George Hunter he first investigated the isolation and bioassay of ascorbic acid in rose hips and other sources. Between 1949 and 1962 he and his assistants published many papers on the activity of serum and intestinal enzymes: alkaline phosphatases, lipases, amylases, and sucrase.

The first course in biochemistry at the University of Alberta was given by Collip in 1915 in the University Power House. It was within the Department of Physiology and Pharmacology under Professor H.H. Moshier. In 1921 the newly formed Department of Biochemistry was located in the Medical Building, and in 1972 it was relocated in the new Medical Sciences Building.

TORONTO

As recorded above, the first university department of biochemistry in Canada was established at the University of Toronto in 1908 with A.B. Macallum as professor until 1917. He was succeeded by T. Brailsford Robertson (1884-1930) (17, 18), an Australian who came to Toronto from the University of California but who returned to Australia in 1919. Andrew Hunter, then professor of pathological chemistry, was appointed to succeed him as head of biochemistry and became one of Canada's more eminent biochemists.

Andrew Hunter (1876-1969) (19) was born and educated in Edinburgh. He was house officer at the Royal Infirmary (1901-2), Crichton Research Fellow in the Department of Physiology at Edinburgh (1902-5), and Carnegie Research Fellow at the Friedrich-Wilhelm University in Berlin with Emil Alderhalden (1905-6) and at the Physiological Institute of the University of Heidelberg with Albrecht Kossel (1906-7). He was appointed assistant professor of biochemistry at Cornell (1908-14) and was associated with the United States Public Health Service (1914-15) as biochemist in charge of the investigation of pellagra. He then became successively professor of pathological chemistry (1915-20) and professor of biochemistry (1919-29) at Toronto. He resigned after nomination to the Gardiner Professorship of Physiological Chemistry at the University, Glasgow (1929-35), and dean of the Faculty of Medicine (1930-35). A cycle was completed when he returned to Toronto in 1935 as professor of pathological chemistry. In 1945 he became dean of graduate studies. After his retirement in 1947 at the age of seventy, he continued scientific work as a research associate at the Hospital for Sick Children in Toronto until his ninetieth birthday, on 28 September 1966. He died on 11 July 1969, in his ninety-third year. During his years at the University of Toronto he had many post-graduate students among whom were W.R. Campbell, D.A. Scott, C.C. Lucas, J.A. Dauphinee, G.S. Eadie, W.D. McFarlane, J.A. Morrell, J.M.R. Beveridge, A.G. Gornall, and C.S. McArthur.

Hunter's early research was concerned with specificity of protein precipitins induced by injection of whole serum or of purified plasma proteins into animals and of precipitins present in snake antivenoms and antisera. With M.H. Givens and others he compared the purine metabolism in a variety of animals. With Walter R. Campbell he made a study of creatine and creatinine which led to the publishing of his classical monograph on the subject, *Creatine and Creatinine*, in 1928. To enable him to use arginase as a tool to estimate arginine, Hunter studied the characteristics of its activity. This led to the determination of its distribution in various organs of a wide variety of vertebrates (with J.A. Dauphinee, q.v.) and in turn to the examination of some aspects of the citrulline-ornithine cycle in the formation of urea (with A.G. Gornall, q.v.). Hunter's last published contribution was an excellent procedure for the estimation of histidine which led to the discovery of a genetic disease in babies, histidinaemia. His last paper was published in *Clin. Chem. Acta* (12: 2 [1965]), sixty-one years after his first. His span of scientific activity was thus comparable to that of D.D. Van Slyke in the United States, E.F. Terroine in France, and F.G. Hopkins and R.A. Peters in England.

Andrew Hunter was an elegant, meticulous experimentalist, a devoted scientist, and an able teacher. He dressed fastidiously and spoke slowly and deliberately, with careful diction. Many students declared 'Andy's' courses to be the most lucid lectures in the whole Faculty of Medicine, based on perfectly organized material given a polished delivery. However, because he seemed to possess little sense of humour and usually appeared austerely correct, Andrew Hunter inspired admiration rather than affection in his students. His seemingly cold personality may have been only Scottish reticence; the few who gained his friendship were astonished at the cordiality of which he was capable away from the academic environment. Andrew Hunter was elected a Fellow of the Royal Society of Canada in 1916 and served as president of Section V in 1924-25, and a Fellow of the Royal Society of Edinburgh in 1932. The CBE was conferred on him in 1946 for his services in the Standing Committee on Nutrition.

Hardolph Wasteneys (1881-1965) (20) succeeded Hunter in 1929 as professor of biochemistry and held the chair until retirement in 1951. He was an Englishman by birth, the eldest son of Sir Wm. Wasteneys, Bart. He went to Australia as a boy and served as a government biologist and chemist, concerned mainly with a study of water purification, principally in Brisbane. To learn more about American procedures in this subject he went to California about 1909, and then found his way to the Rockefeller Institute in New York where he worked for several years (1910-16) as an associate in the laboratory of Jacques Loeb, with whom he performed the early classical experiments on aspects of cellular oxidation and artificial parthenogenesis. He was awarded a PHD by Columbia University in 1916, without any previous academic degree, and then moved to the University of California for a year. In 1917, with T.B. Robertson, he went to Toronto as assistant professor of biochemistry, where he remained until his death. His major scientific interest was enzymology, and, with Henry Borsook, he published a series of notable papers in 1924 on the synthesis of 'plastein' with pepsin. His later published papers were relatively few.

Wasteneys had a broad humanitarian outlook and a multitude of interests. He was an enthusiastic student of history and, for many years, gave a series of popular lectures on the history of science and civilization. He was deeply interested in the University Settlement and in the operation of Hart House as a member of the governing body. In gratitude his portrait, painted by F.H. Varley, hangs in the latter building. Wasteneys was elected a Fellow of the Royal Society of Canada in 1930 and served as president of Section V (1940-41).

While Wasteneys was head of the department a succession of young biochemists from England were appointed to the staff to direct graduate work. Thus H.D. Kay (1929-32) published papers on phosphoric esters and their hydrolysis by phosphatases, Guy F. Marrian (1933-38) on steroids and sex hormones, and Leslie Young (1939-47) on mechanisms of detoxication. From 1947 to 1957 G.C. Butler (q.v.) did notable early work on the preparation and structure of nucleic acids and nucleoproteins. Among their more distinguished post-graduate students were W.R. Graham Jr, H.D. Branion, and T.H. Jukes (with H.D. Kay); G.C. Butler, S.L. Cohen, A.D. Odell, and D. Beall (with G.F. Marrian); D.H. Laughland, J.A. McCarter, and S.H. Zbarsky (with L. Young); C.W. Helleiner, R.O. Hurst, A.M. Marko, L.B. Smillie, D.B. Smith, and J.M. Neelin (with G.C. Butler); and H. Borsook, H.B. Collier, C.A. Morrell, J. Campbell, B.F. Crocker, and A.R.G. Emslie (with H. Wasteneys).

A small Department of Enzymology was formed at the University of Toronto in 1919 for H.B. Speakman (1903-75) who had come over from England in 1916 during World War I with bacterial cultures for the industrial production of acetone by the fermentation of cereal starches. Acetone was urgently required at that time as a solvent. When Speakman became head of the Ontario Research Foundation in 1929 he was succeeded by A.M. Wynne and the Department of Enzymology was absorbed into the Department of Biochemistry.

Arthur Marshall Wynne (1892-1972) was born in Brigden, Ontario, in 1892. He was educated initially at Queen's University, and then at Toronto, where he took a PHD in 1925. He was a research associate in enzymology with H.B. Speakman from 1919 until 1929, and then joined the staff of the Department of Biochemistry, rising from associate professor to become its head from 1951 until retirement in 1960. He published relatively few papers but was a conscientious teacher and administrator. Among the twenty-one students who qualified for the PHD under him were R.D.H. Heard, W.W. Johnston, and L.B. Pett. His publications were concerned mainly with parameters of activity of pancreatic and bacterial enzymes. He was elected as the first president of the Canadian Biochemical Society (1957-58), and to Fellowship in the Royal Society of Canada in 1943. Like Wasteneys he had a major external interest – music – and he was a gifted pianist and organist.

Wynne was succeeded as head of the Department of Biochemistry by Charles S. Hanes in 1961. Charles Hanes was born in Toronto in 1903. He obtained his BA from Toronto University in 1925, lectured at Queen's University from 1927 to 1928, and then went to England for his PHD at Cambridge University. Upon his return to Canada he was appointed to the staff of the Ontario Research Foundation in 1932. In 1934 he returned to England to the Low Temperature Station at Cambridge, where from 1934 to 1947 he carried out notable investigations with amylases on the structure of starch. He discovered amylose phosphorylase, first postulated the possible existence of a helical conformation in a macromolecule in 1937, and accomplished the synthesis of starch from potassium glucose-1-phosphate in 1940. Hanes served as director of the Agricultural Research Council unit from 1947 to 1951, when he returned to Canada as professor of biochemistry at Toronto. He became head of the department in 1961, with emeritus status from 1965. During this period Hanes investigated enzymic reactions related to peptides, and quantitative chromatographic methods for amino acids, peptides, and nucleotides. He was elected a Fellow of the Royal Society of London in 1942 and of Canada in 1956, and was the Flavelle medallist in 1955.

George Edward Connell succeeded Hanes as head of the department in 1965. He was born in Saskatoon in 1930. He graduated from the University of Toronto in 1955 with a doctorate in biochemistry under the direction of Hanes and joined the staff there in 1957. Since then he and his students have published numerous papers on various aspects of protein chemistry and in particular the haptoglobins. He has studied enzymic transpeptidation and the metabolism of bacterial peptides. He became associate dean of medicine in 1970 and was succeeded as head of biochemistry by George Ronald Williams, who had come over from England in 1952 to join the staff of the Banting and Best Department of Medical Research and had transferred to the Department of Biochemistry as associate professor in 1961.

The Old Medical Building was the site of the first instruction in biochemistry at the University of Toronto about 1900. This location continued to house the department, when formed, until the opening of the new Medical Sciences Building in July 1968.

Laurence Irving, when at Toronto as professor of experimental biology (1927-37), inspired numerous fundamental studies of the comparative physiology of respiration and growth, especially in aquatic animals. Irving went to Swarthmore and later became scientific director of the Arctic Research Laboratory, Anchorage, Alaska. As one of his post-graduate students, Jeanne Manery (Fisher) began a series of investigations into the role of electrolytes and ionic transport in the cellular metabolism of piscine and amphibian muscles. These she has continued since 1939 in the Department of Biochemistry.

The chemistry of food and dietetics have formed part of college curricula for many years, the former as a course in departments of chemistry and the latter in schools of household science. In fact at the University of Toronto a degree course in domestic science was established in 1902 in which Clara Cythia Benson (1875-1964) taught courses on food chemistry, after her graduation with a PHD in physical chemistry in 1903. She then became a member of A.B. Macallum's staff and taught in the new Faculty of Household Science. In 1919 the university created a separate department of food chemistry and Clara Benson was its head from 1919 until retirement in 1945. She was succeeded by Doreen Smith. Few original contributions of a biochemical nature have been published from the department.

With the discovery of vitamins and their relation to deficiency diseases the discipline of nutrition was born in the early part of the twentieth century. It is natural that the subject should have been taught primarily in a school of public health apart from its place in the teaching of biochemistry, physiology, pharmacology, general medicine, and in agricultural subjects. A Department of Nutrition was established in the School of Hygiene at the University of Toronto in 1946 with E.W. McHenry as head and it was the only school of its kind in Canada for many years.

Earle Willard McHenry (1899-1961) (21) was born in Streetsville, Ontario, on 25 January 1899. He received his BA with honours in chemistry from the University of Toronto in 1921 and his MA in 1923. The topic of his thesis, 'Studies on the Composition of Saliva,' reveals an early interest in the biological applications of chemistry. He gave lectures in chemistry in the Faculty of Dentistry during 1921-23, and then worked in the laboratories of Canadian Canners Ltd, in Brighton, Ontario (1923-25). After two years with E.R. Squibb and Sons, in their biological laboratory in New Jersey, McHenry returned to Toronto in 1927 as an assistant in the newly formed Department of Physiological Hygiene. He was concerned with the production of glandular extracts in the Connaught Laboratories and developed a procedure for the preparation of an active liver fraction for oral use in the treatment of pernicious anaemia and later one for intramuscular injection. He received his PHD in 1929 for a thesis on the chemistry of histamine. McHenry was appointed lecturer in the School of Hygiene and gave a course in nutrition for students in the diploma course in public health. In 1935 the first MA degree in nutrition was conferred and in 1938 the first PHD under his supervision.

McHenry had joined C.H. Best and his colleagues in their study of fatty livers caused by a dietary lack of choline and its precursors, the so-called lipotropic agents. For more than ten years aspects of this nutritional problem, and others concerned with fat metabolism, were his major research interests. In 1941, with Gertrude Gavin, he discovered the lipotropic action of inositol on a fat-free diet. McHenry studied the effects of vitamin C, thiamine, pyridoxin, and biotin on fat metabolism. During thirty-four years in the School of Hygiene he and his associates published more than 150 papers in nutrition.

Teaching nutrition was also of great interest to him. In 1935 he was appointed assistant professor and in 1938 associate professor of physiological hygiene. In 1942 a sub-department of nutrition was created and he was placed in charge. In 1946, it became a full department with McHenry as professor of public health nutrition in the School of Hygiene. In 1959 the name was changed to the Department of Nutrition.

McHenry played a leading role in the formation of the Canadian Council of Nutrition in 1937, and served on it until he died. He was active also in the group that formulated the Canadian Dietary Standards. By public lectures, radio addresses, and later on television he tried to educate the public to adopt sensible food habits. McHenry was an excellent public speaker and stimulating teacher. His enthusiasm was infectious. He conducted several dietary surveys because of interest in the nutrition of Canadians and served on several governmental committees, both national and provincial. Twice he attended the International Congress on Nutrition as a Canadian delegate.

He was instrumental in bringing about the formation of the Canadian Federation of Biological Societies in 1957 and the Nutrition Society of Canada in 1958, serving as first president of both organizations. He was elected a Fellow of the Royal Society of Canada in 1942.

McHenry was the author of two books, *Basic Nutrition*, a college textbook, published in 1957, and *Foods Without Fads*, a popular guide, in 1959. At the time of his death he was acting as co-editor with G.H. Beaton of *Nutrition: A Comprehensive Treatise* which was published in three volumes in 1964-65.

McHenry may be considered as the founder of the first school of modern nutrition in Canada. He was succeeded by one of his former post-graduate students, George Hector Beaton, who took his PHD in nutrition in 1955 and was then appointed to the staff as assistant professor in the department. He was promoted to full professorship and head of the department in 1963. In 1961 he was in Central America and Panama as a Fellow in the service of the World Health Organization. His investigations concern nutritional aspects of pregnancy. He was co-editor with McHenry and largely responsible for the publication of *Nutrition: A Comprehensive Treatise*. He served for several years on the Canadian Council on Nutrition and was president of the Nutrition Society of Canada (1965-66).

MANITOBA

Alexander Thomas Cameron (q.v.) began the teaching of biochemistry in 1909 in the Department of Physiology and was appointed head of an independent Department of Biochemistry in 1923. He held the chair until his death in 1947 at the age of sixty-five. His senior assistant, Frank David White (1892-1970), a Glaswegian who had come to Canada in 1924, succeeded him and continued as head of the department until retirement in 1959. He was succeeded by Paul Beo Hagen, an Australian of broad experience from Sydney, who resigned in 1964 to assume the professorship at Queen's. He was succeeded by Marcel Corneille Blanchaer, a medical graduate from Queen's University who had previously been head of the biochemical laboratory of St Boniface Hospital and on the staff of the medical school in Winnipeg.

George Edward Delory, a student of E.J. King at the British Postgraduate Medical School in London from 1935 to 1941, joined the staff as assistant professor in 1948. He retired because of ill health in 1964. Delory published a small manual entitled *Photoelectric Methods in Clinical Biochemistry* in 1949 and contributed valuable data and many useful procedures, modifications, and inventions in this field. In the department, for the past ten or more years, Ester Yamada has been investigating the enzymic oxidation of aldehydes, the metabolism of nucleic acids, especially in bone marrow, and the role of uridine phosphorylases therein.

Biochemistry was taught in the old medical building in Winnipeg from 1909 until 1973 when the new Basic Medical Sciences building was opened.

WESTERN ONTARIO

With the reorganization of the Faculty of Medicine at the then Western University, E. Gordon Young (q.v.) first taught systematic biochemistry there in 1921. On his appointment to Dalhousie in 1923 he was succeeded by Archibald Bruce Macallum (1885-), a son of A.B. Macallum and a medical graduate of the University of Toronto, who remained as head of the department until 1947. After graduation in 1909 Macallum had worked with O. Folin at Harvard on the estimation of uric acid (1909-12) and then with A. Harden and C. Funk at the Lister Institute in London as a Beit Memorial Fellow (1913-14) on the isolation of vitamin B. He then taught biochemistry at Toronto from 1914 to 1916, after which he went into commercial work with the Synthetic Drug Company before his move to Western. At Western he served as dean of medicine from 1927 until retirement in 1954.

Macallum was followed by Roger James Rossiter, an Australian by birth and a Rhodes Scholar at Oxford where he began research under Sir Rudolph Peters. At the University of Western Ontario Rossiter and his colleagues carried on extensive investigations of the biosynthesis of the phospholipids in brain and nerves for a period of twenty years and developed an international school of neurochemistry. He was elected to the Royal Society of Canada in 1954 and was Flavelle medallist in 1963. When Rossiter became dean of the graduate school in 1965 he was succeeded by Harold Brown Stewart as head of the department, and in 1971 when Rossiter became vice-president (academic) Stewart again succeeded him, as dean.

Kenneth Percy Strickland should be mentioned here for his contributions to the biosynthesis of lecithin and the role of cytidine diphosphate choline therein. Previously, with R.H.S. Thompson at Guy's Hospital Medical School, he also studied the incorporation of P^{32} in brain slices and the action of phosphatases on lipid phosphoric esters. Strickland graduated from the University of Western Ontario and took his doctorate in 1953 under Rossiter. He then spent two years at Guy's Hospital in London as a Fellow of the National Research Council. Subsequently he was appointed to the staff at Western and promoted to full professorship in 1966 and acting head in 1972.

Biochemistry was taught in the old medical building opposite the Victoria Hospital from 1921 until 1965, when the department was moved into the new medical complex on the university campus – on the site of the old golf course of the London Hunt Club, where the Collip Research Laboratory was built initially.

DALHOUSIE

Systematic biochemistry was first taught at Dalhousie and a department established in 1924 with Elrid Gordon Young as professor, by virtue of a grant from the Rockefeller Foundation. Young was a graduate of McGill under R.F. Ruttan and V.J. Harding and of Cambridge under F.G. Hopkins. Between 1924 and 1950 he and his students investigated the proteins of eggs; they were the first in Canada to apply the Tiselius electrophoretic method to the analysis of the proteins in eggs and in human sera of pathological states. Other investigations were made in the fields of nutrition, toxicological chemistry, purine metabolism, especially allantoin, and the chemical composition of marine algae. His post-graduate students included G.A. Grant, M.R. Butler, R.W. Begg, W.W. Hawkins, R.V. Webber, W.R. Inman, F.A.H. Rice, C.F.C. MacPherson, J.B. Neilands, and D.W. Watson. In 1950 Young resigned to become director of the new Atlantic Regional Laboratory of the National Research Council in Halifax, where he remained until retirement in 1962.

At Dalhousie Young was succeeded by John Alexander McCarter, who held the professorship until 1965. McCarter was born in England in 1918. He was educated at the University of British Columbia and took his PHD at Toronto in 1945 under Leslie Young. After three years of research at Chalk River, Ontario, as an assistant research officer of the National Research Council, he was appointed an associate professor of biochemistry at Dalhousie University in 1948. From 1950 to 1965 he was professor and head of the department. McCarter resigned to become director of the Cancer Research Unit at the University of Western Ontario. His interest in research for many years has been in experimental carcinogenesis. He was elected a Fellow of the Royal Society of Canada in 1964.

Christopher Walter Helleiner was appointed to succeed McCarter as head of the department. Although born in Vienna in 1930 he had been educated at the University of Toronto, where he took a PHD in biochemistry in 1955 under G.C. Butler. As a Fellow of the National Research Council he spent two years at Oxford in microbiology with D.D. Woods. He then was successively an assistant scientist with the Ontario Cancer Institute (1957-59), assistant professor of medical biophysics at the University of Toronto (1959-63), associate professor of biochemistry at Dalhousie and head of the department in 1965. His main subject of research has been the biosynthesis of nucleic acid in virus-infected animals.

Stanley D. Wainwright, previously with Atomic Energy of Canada Ltd, was appointed to the staff in biochemistry as a research associate of the Medical Research Council in 1956 and promoted to full professor in 1964. He has investigated several fundamental aspects of the biosynthesis of proteins in *Neurospora crassa* and of haemoglobin in the blastodiscs of embryonic chicks. In 1972 he published a book entitled *Control Mechanisms and Protein Syntheses*.

In 1960 Sydney John Patrick joined the department from Jamaica and brought with him the problem of the hypoglycaemic action of Hypoglycin A (β-methylene cyclopropylalanine), an unusual amino acid found in the seeds of *Blighia sapida*, the ackee plant. This he has shown to be due to its inhibition of hepatic and renal gluconeogenesis with damage to mitochondria, decreased ATP, and decreased phosphorylation of glucose and fructose. These studies have been extended to the action of Phenformin (phenethyl diguanide) as another hypoglycaemic agent.

Lloyd Bertram Macpherson was appointed dean of the Faculty of Medicine at Dalhousie in 1971, an unusual distinction for a biochemist without medical qualification. A Maritimer by birth, he took his doctorate in pathological chemistry in 1949 at Toronto under Andrew Hunter, but working with C.C. Lucas, with war service in the army intervening. After six years at the Banting Institute Macpherson joined the staff at Dalhousie in the Department of Biochemistry.

He was promoted to full professorship in 1963 and made assistant dean of medicine in 1958.

The Department of Biochemistry was located initially in the Medical Sciences Building when it was constructed in 1923 and remained there until the opening of the fifteen-storey Sir Charles Tupper Building in 1967.

Mention should be made here of the pioneering investigations of Frederick Ronald Hayes in the Department of Biology between 1930 and 1953 on the chemical embryology of fish eggs. Fifteen articles were published on the chemical composition and metabolism of salmonid eggs during development. These studies demonstrated the rates of oxidation and translocation of protein from yolk to embryo, the steady rise of non-protein nitrogen in the embryo, the rapid disappearance of fat to provide energy especially between twenty and forty days after hatching, the enzymic mechanism of hatching, and the requirement of oxygen and some physical parameters in development. Hayes was the director of the Dalhousie Institute of Oceanography (1959-64) and then chairman of the Fisheries Research Board of Canada until his retirement in 1969.

Special mention should be made of Leo Charles Vining, who was elected a Fellow of the Royal Society of Canada in 1974 and who has made notable contributions to the biosynthetic pathways of antibiotics and to fungal metabolism. He was born in Whangarei, New Zealand, in 1925 and acquired his initial university education in that country at the University of Auckland. He was awarded an 1851 Exhibition Scholarship which permitted him to qualify for the PHD in organic chemistry in 1951 at Cambridge University, where he worked under Lord Todd. Vining then spent a year with S. Waksman at Rutgers University in New York. He was appointed to the staff of the National Research Council of Canada, initially at the Prairie Regional Laboratory (1955-62) and latterly at the Atlantic Regional Laboratory (1962-71), where he rose to the rank of principal research officer. In 1971 he accepted a professorship in the Department of Biology at Dalhousie University. Vining's original interest was in the chemical structure of antibiotics and other metabolic products of fungi but in recent years he has studied their biosynthetic pathways.

MONTREAL

At the University of Montreal Georges H. Baril (1885-1953) was appointed to teach chemistry in 1911. He was a graduate of that University (BA, 1904) and of Laval (MD, 1908). His teaching was to medical and dental students but from 1920 he was associated with the Department of Chemistry in the Faculty of Science in which he was a professor of physiological chemistry until a separate Department of Biochemistry was created in the Faculty of Medicine in 1951. A pupil of his, Jules Labarre, was responsible for the teaching of biochemistry in the Faculty of Science from 1929 until 1945. Baril also taught organic chemistry to pre-medical students. He did very little research but trained several graduate students who later became professional biochemists: J. Labarre, G. Gosselin, and L.-P. Bouthillier. In several respects he played a role comparable to that of Ruttan at McGill. He retired in 1952 and died in 1953.

Louis-Phillippe Bouthillier joined the staff of the Department of Chemistry in 1944 and assisted in the teaching of biochemistry to science students. When Labarre devoted all his time to the Faculty of Pharmacy in 1945, Bouthillier took over all the teaching in the department until he joined the new Department of Biochemistry in 1951. After the resignation of Baril in 1952 Bouthillier took over from him and carried on the department until 1964 when Walter G. Verly from the University of Liège was appointed to the chair.

No biochemistry was taught officially in the Faculty of Science between 1951 and 1955. Then it was assumed by Edouard Pagé in the Department of Biological Sciences, and an honours course instituted. However, in 1965 responsibility for teaching honours and graduate students reverted to the Department of Biochemistry in the Faculty of Medicine, subject to the approval of courses by the Faculty of Science. It is thus still a complex and confusing situation at the University of Montreal.

Walter G. Verly was born in Péronnes, Belgium, in 1923 and took his doctorate in medicine at the University of Liège in 1947. After holding fellowships in biochemistry at Edinburgh (1948-49) and at Cornell (1949-51), he was appointed an assistant professor at Liège in 1952, where he remained until his appointment as head of the department at the University of Montreal in 1964. His fields of investigation are intermediate metabolism, catecholamines, and molecular genetics.

A graduate student under Bouthillier, Robert Gianetto, joined the staff in 1953 as associate professor after working on enzymes in mitochondria for a year with C. De Duve at the Catholic University in Louvain. Subsequently he has investigated β-glucuronidase in hepatic cells and the biosynthesis of glucuronides. Jean Paul Lachance, a graduate of Laval, was appointed to the staff in 1958 after spending a year with G.J. Popjak at the Medical Research Council hospital at Hammersmith as a Nuffield Fellow. There he became interested in the biosynthesis of fatty acids and has carried on research in this field subsequently.

Mention should be made here of the many contributions of Hans Selye as director of the Institute of Experimental Medicine and Surgery at the University of Montreal since 1945. Born in Vienna in 1907 and educated at the Prague Technical University (MD, 1929; PHD, 1931) he came to North America in 1931 to the Johns Hopkins Medical School. In 1932 he moved to McGill as a Rockefeller Fellow, where he was associated with Collip for several years before his appointment to his present position at the University of Montreal. His field of research in biochemistry has been endocrinology, especially of steroids in the physiological state of stress and in cardiovascular diseases. He has published extensively and was elected a Fellow of the Royal Society of Canada in 1941.

LAVAL

The teaching of biochemistry was shared by two faculties at Laval initially. It began in the Faculty of Science in 1928 with Elphège Bois as professor, and from 1929 to 1935 he was also responsible for teaching the subject in the Faculty of Medicine as well. However, J. Rosaire Gingras assumed the responsibility for the course in the medical faculty from 1935 and served as professor of the new department from its inception in 1940 until 1963. Gingras was succeeded by Louis Berlinguet as professor of biochemistry in the Faculty of Medicine.

Berlinguet was born in Three Rivers, Quebec, in 1926. He took a baccalaureate in science at the University of Montreal and the PHD at Laval in chemistry. He was appointed to the staff of the Department of Biochemistry at Laval in 1947, and became full professor in 1961 and head of the department in 1963. Berlinguet deserves credit for developing a field of investigation of the synthesis and metabolic effects of unnatural amino acids, begun in the Department of Chemistry by Roger Gaudry (q.v.), whose post-graduate students included L. Berlinguet, C. Gaudin, G. Nadeau, K.F. Keirstead, and F. Martel. Berlinguet was elected a Fellow of the Royal Society of Canada in 1969. On his appointment as vice-president (research) of the newly established University of Quebec that same year, he was succeeded by L.-M. Babineau at Laval.

The Department of Biochemistry in the Faculty of Science was established in 1943. Bois served as its head until his retirement in 1961. He was succeeded by Marcel Jean, a local graduate, and he in turn by Patrick Tailleur in 1969. Within the past decade the department has grown rapidly.

Claude Godin joined the staff in 1961. He had earned the DSC from Laval in 1953, received a fellowship (1953-55), and taught at the University of Ottawa from 1956 to 1961 as assistant professor. His research has been in the field of proteins: the end-groups of keratin, biosynthesis in the lactating goat, metabolism of phenylalanine and tyrosine in the rat, and reactivity of pepsin to substrates with C^{14} as indicator.

Fernand Labrie, also a graduate of Laval, was appointed director of the Laboratory for Molecular Endocrinology in 1969 after spending three years in England, two of which were in Cambridge with F. Sanger. There he isolated the m-RNA of haemoglobin. He has published several articles on the enzymes and humoral releasing mechanisms of endocrine glands.

M.R.V. Murthy, a native of India and a graduate of Bangalore, was appointed to the staff in the Faculty of Medicine in 1964 and became full professor in 1969. Since his arrival in Canada he has been investigating the metabolism of ribosomes and of phenylalanine in brain tissue.

Biochemistry was first taught in the old building of the Medical School built in 1854 in the Quartier Latin behind the seminary and near the ramparts overlooking Mountain Hill and the Hotel Dieu. Only in 1957 did the Chemistry Department move to the new Pavillon Vandry on the campus on Ste Foy Road, now known as the Cité Universitaire.

QUEEN'S

At Queen's University biochemistry was first taught in the Department of Chemistry from 1914 to 1925 by A.P. Lothrop (1884-1944), a pupil of W.J. Gies of Columbia, and by J.F. Logan (1884-1951) from 1925 to 1950 (22). Financed by a donation from Dr Agnes Craine, a separate department was established in 1937 in the Craine Building with Robert Gordon Sinclair (1903-49) (23), a pupil of W.R. Bloor at Rochester and a graduate of Queen's, as professor. Sinclair made excellent fundamental studies of the properties and metabolism of the phospholipids and triglycerides in animals, with elaidic acid as tracer. With Malcolm Brown he was a co-leader of the first Arctic expedition of Queen's University in 1947 and was a passenger on the ss *Nascopie* when it foundered. He analysed many samples brought from Southampton Island subsequently. He was an active member of the Canadian Physiological Society and its secretary for five years. Sinclair was elected to Fellowship in the Royal Society of Canada in 1943.

After his untimely death by drowning in 1949 Sinclair was succeeded by James MacDonald Richardson Beveridge. Beveridge was born in Dunfermline, Scotland, in 1912, and came to Canada with his family in 1927. He obtained his BSC from Acadia in 1937 while earning his keep during the summer by cutting the acres of grass on the campus. Beveridge went to Toronto to work in Banting's laboratory with C.C. Lucas, and registered as a graduate student in the Department of Pathological Chemistry. He obtained his PHD from Toronto in 1940 and remained with Lucas as research associate until 1944. Work on the analysis of proteins and some studies of lipotropic phenomena prepared him for a position at the Pacific Experimental Station of the Fisheries Research Board, where he did comparative studies of the nutritive value of the flesh of a number of fishes with that of several animals (1944-46). Beveridge then lectured in pathological chemistry at the University of Western Ontario (1946-50), during which time he acquired an MD. In 1954 he was appointed professor and head of the Department of Biochemistry at Queen's University, where he stayed until he accepted the presidency of Acadia University in 1964. At Queen's he was chairman of the Board of Graduate Studies (1960-63) and dean of the School of Graduate Studies (1963-64). Beveridge was also chairman of the Defence Research Board Panel on Nutrition and Metabolism (1961-65). In spite of all these commitments, before he left the laboratory for the president's office he and his students had published about 100 scientific papers on such subjects as protein analysis, dietary and hepatic necrosis, and the effect of diet, especially the fat component, on lipaemia. The latter studies comprise an experimental approach to cholesterolaemia, atherosclerosis, and related coronary heart disease. Beveridge was elected a Fellow of the Royal Society of Canada in 1960.

Beveridge was succeeded at Queen's by Paul Beo Hagen (q.v.) in 1964, who became vice-chairman of the Medical Research Council and in 1968 professor and dean of Graduate Studies at the University of Ottawa.

Peter Harry Jellinck succeeded Hagen at Queen's. Jellinck has had a wide experience both in England and Canada. Although born in Paris, France, in 1928, he was educated in England, where he took his PHD in biochemistry at the University of London in 1954. He came to McGill for one year as a Post-doctorate Fellow of the National Research Council, then returned to England to teach chemistry successively at the Norwood Technical College (1955-57), St Bartholomew's Hospital Medical School (1957-58), and the Middlesex Hospital Medical School (1958-59). He was appointed associate professor of biochemistry at the University of British Columbia in 1960 and rose to full professorship in 1967. He has investigated the relationship of chemical carcinogens to steroid metabolism and aspects of the chemistry and metabolism of oestrogens. In 1963 Jellinck published a textbook entitled *Biochemistry: An Introduction*.

Another member of the department, with professorial status since 1968, is Robert Osmund Hurst. He took his PHD in 1952 under Butler at Toronto, where he became interested in nucleic acids. Subsequently he has published studies of the enzymic and alkaline hydrolysis of calf thymus DNA and the products formed. More recently he has investigated computer programming of steady-state and non-steady-state rate equations in enzymic reactions.

The investigations of J.K.N. Jones, in the Department of Chemistry at Queen's, are of biochemical importance in the field of carbohydrates. In 1953 Jones came to Canada from the University of Bristol to be Chown Research Professor at Queen's, where he has established an international school of carbohydrate chemistry. He had been a student of W.N. Haworth at Birmingham. He was elected a Fellow of the Royal Society of London in 1957 and of Canada in 1960. He and his students have investigated the chemical composition, structure, and biogenesis of many plant carbohydrates, especially hemicelluloses. Within recent years Jones has been studying the biosynthesis of acetamido-deoxyketones from the corresponding hexitols and pentatols by *Acetobacter suboxydans*.

34 The development of biochemistry in Canada

SASKATCHEWAN

At the University of Saskatchewan Rodger J. Manning (1883-) began teaching biochemistry in 1916 in the Department of Chemistry. He had been graduated from Toronto as an inorganic chemist and taught chemistry for six years at Toronto and Queen's. He went to England and received the DSC degree after several years of research at the University of Bristol with M. Nierenstein on chemical aspects of tannins. Subsequently he spent some time in Germany at Munich and Greifswald in the fields of physiology and biochemistry and started work for the PHD with O. Dimroth. World War I interrupted and he returned to Toronto as lecturer in biochemistry with Macallum. There he completed also all of the courses required for the MB but never the internship. On Macallum's advice he moved to Saskatoon in 1916, where he taught biochemistry for thirty years. He resigned at the age of sixty when the new medical building was constructed and a transfer of location was necessary. Manning did little research after graduation and was devoted to music, especially vocal. He sang frequently and conducted local choral groups. He remained a bachelor all his life. On his retirement in 1946 a separate Department of Biochemistry was established with Herbert Bruce Collier as professor.

Collier had been trained as a biochemist under Wasteneys at Toronto and had subsequently spent seven years teaching at the West China Union College of Medicine (1932-39). He then worked as biochemist at the Institute of Parasitology (Macdonald College) from 1939 to 1942 and taught at Dalhousie from 1942 to 1946. He resigned from Saskatchewan in 1949 to become head of the department at Alberta. Collier was succeeded by Charles Stewart McArthur, who had been trained under Andrew Hunter at Toronto. He in turn was followed in 1968 by James Douglas Wood, a Scot from Aberdeen who emigrated to Canada after graduation from the university there and who had been employed as a research worker in the Department of Agriculture (1954-57), the Fisheries Research Board (1957-61), and the Defence Research Board (1961-68).

Another biochemist at the university is Robert William Begg, a Maritimer by birth and education, who graduated from Dalhousie with the MSC in biochemistry in 1938 and MD in 1942. He became one of the few paratroopers in the RCAMC during World War II and, on demobilization, qualified for the DPHIL (OXON) in pathology under Sir Howard Florey. He returned to Canada as assistant professor of biochemistry (1946-48) at Dalhousie, where he developed a unit of the National Cancer Institute (1948-50). In 1950 he moved to London, Ontario, as associate professor of medical research of the National Cancer Institute in the Collip Medical Research Laboratories, where he stayed until 1957. Begg then came to the University of Saskatchewan as head of a unit of the National Cancer Institute (1956-62). He was dean of Medicine from 1962 and professor of chemical pathology from 1964 until 1968, when he became vice-president of the university (principal of the Saskatoon campus). His scientific contributions have been on enzymic aspects of cancer metabolism and systemic effects of malignancy. With his collaborators he established a decrease in the activity of liver catalase and of biosynthesis in adipose tissue from acetate in tumour-bearing rats. Begg was editor of the proceedings of five Canadian Cancer Research Conferences held at Honey Harbour, Ontario (1958-62).

OTTAWA

The Medical School was opened in 1945 but the early years of its Department of Biochemistry were difficult ones because of unsuitable conditions and temporary appointments: V. Vlassopoulos (1946-47), D. Hingerty (1947-48), M. Murnaghan (1948-49), and J. Ettori (1949-61). Practically no research was done there before 1961 when Antoine D'Iorio was appointed professor and head of the department. Bernard Roland Belleau, professor of biochemistry in the Department of Chemistry, taught the subject from 1958 to 1969 and carried on active research in the fields of alkaloids and the metabolism of drugs.

D'Iorio was born in Montreal in 1925. After graduating from the University of Montreal with the doctorate in 1949, he taught physiology there from 1949 until 1961, except for one year at the University of Wisconsin and one at Oxford. He then moved to Ottawa. With his associates D'Iorio has made important contributions over twenty years to the metabolism of the adrenal gland and its chromaffin granules. He established that catecholamines exist as a salt of ATP bound to protein. He has shown that adrenolytic substances release catecholamines from the adrenal and that the thyroid hormones have an inhibitory effect on amine oxidases and O-methyl transferase which he has isolated and characterized. In 1969 D'Iorio was appointed dean of the Faculty of Pure and Applied Science and the teaching of biochemistry was consolidated in one department serving both the faculties of science and medicine. D'Iorio was elected a Fellow of the Royal Society of Canada in 1969. He was succeeded as head of the department by Donald Sainteval Layne.

Layne was born in Lime Ridge, Quebec, in 1931. He took his PHD at McGill in 1957 and then went to Edinburgh as a Fellow in biochemistry for one year. In 1958-59 he was a research associate in psychiatry at Queen's, and then a scientist at the Worcester Foundation of Experimental Biology (1959-64) and at the Food and Drug Directorate in Ottawa (1964-68). He succeeded D'Iorio as professor of biochemistry at the University of Ottawa in 1969. His research interests have been oestrogenic hormones, initially the steroids of birds, then the pathway of oestrogen metabolism, and latterly contraceptive steroids.

Normand Leo Benoiton joined the staff as assistant professor in 1961. He had been educated in biochemistry at the University of Montreal under Bouthillier and then spent two years at the National Institute of Health in Washington and two years at the University of Exeter as ICI Fellow, where he became interested in the methods of synthesizing amino acids and peptides. He has contributed to the synthesis of hydroxyglutamic and hydroxyaspartic acids and acetylserine peptides as related to the active centre of chymotrypsin.

Pierre Proulx joined the staff in 1963. He had graduated from McGill in 1962 and been a lecturer there. He has published papers on the amino acids in the lipids of mammalian brain and the metabolism of phospholipids in *Escherichia coli*.

The staff has been greatly strengthened recently by the appointment as full professors of M. Kates and D.R. Whitaker, both previously with the National Research Council in Ottawa. Morris Kates was born in Rumania in 1923 but was educated at the University of Toronto where he was awarded the PH D in biochemistry in 1948. From 1949 to 1969 he did notable research work on the chemistry, biosynthesis, and metabolism of lipids in plants and bacteria. Recently he has discovered a new class of lipids derived from a phytanyl diester of glycerol present in the membranes of extremely halophilic bacteria. He moved to the University of Ottawa in 1969 and was elected a Fellow of the Royal Society of Canada in 1972. He published a book entitled *Techniques in Lipidology* in 1973.

Donald Robert Whitaker was born in Winnipeg in 1919. He is an enzymologist who, after post-graduate study at the University of London, carried out notable research work from 1948 to 1971 in the Division of Biosciences, National Research Council, Ottawa, on the structure, function, and biosynthesis of microbial, extracellular enzymes, especially proteases and cellulases. In 1971 he joined the staff of the University of Ottawa.

Mention should be made of the fundamental contributions of Keith J. Laidler in the Department of Chemistry. Laidler was graduated from Oxford in 1937 and, as a Commonwealth Fellow, obtained his doctorate at Princeton under Hugh S. Taylor in the field of catalysis. Laidler joined the staff at the University of Ottawa in 1955 and has published, in collaboration with his students, many articles on the kinetics of enzymic catalysis. He published a book on *The Chemical Kinetics of Enzyme Action* in 1958, and was elected a Fellow of the Royal Society of Canada in 1960.

BRITISH COLUMBIA

Failure of the Department of Chemistry at the university to provide a course in biochemistry led to the organization of lectures in this subject in the Department of Poultry Science by J. Biely in 1927 and in the Department of Dairying by B.A. Eagles in 1929. In 1931 W.J. Allardyce started to teach the subject in the Department of Chemistry. Departmental conflicts, however, ended by the transfer of Allardyce to an assistant professorship in the Department of Biology in 1938.

Jacob Biely was born in Russia in 1903. He qualified for the BSA degree in 1926 at the University of British Columbia, the MS at Kansas State College in 1929, and the MSA at the University of British Columbia in 1930. He worked as a research assistant in the Department of Poultry Science until 1942, then was promoted successively until he became head of the department in 1952. He retired in 1968. For almost fifty years he has carried on research in the nutrition of poultry: the nutritive value of animal protein concentrates, the vitamin and amino acid requirements of the growing chick, the utilization of wheat by growing and laying birds, and effects of environment on the production of eggs. He was elected to Fellowship in the Royal Society of Canada in 1966.

Blythe Alfred Eagles was born in New Westminster, British Columbia, in 1902 and educated at the University of British Columbia and the University of Toronto, where he was awarded the PHD under V.J. Harding in the Department of Pathological Chemistry in 1926. He studied the metabolism of creatine and uric acid and was associated with George Hunter in the work on ergotheonine. Eagles was Sterling Fellow at Yale University (1926–28), where he worked with T.B. Johnston, and then spent a year at the National Institute for Medical Research in London. He returned to his Alma Mater in 1929 as assistant professor of dairying and progressed to the head of the department in 1936, where he remained until retirement in 1967. From 1947 to 1967 he was dean of the Faculty of Agricultural Sciences and from 1955 to 1967 chairman of the Division of Animal Science. His investigations have been in the field of bacterial nutrition and the ripening of cheese. He was elected to Fellowship in the Royal Society of Canada in 1952.

William John Allardyce was born in Winnipeg in 1897. He graduated in chemistry from the University of British Columbia (BA, 1919; MA, 1921) and became responsible for the analytical laboratory work as instructor from 1921 until 1929 when he went to McGill on a scholarship. He received his PHD in biochemistry in 1931 under Collip, then returned to his Alma Mater as assistant professor to teach biochemistry in the Department of Chemistry. He resigned after only one year and taught in high school from 1932 to 1938. He again returned to his Alma Mater in 1938 as associate professor in the Department of Biology and was promoted to full professor in 1945. He wrote a paper on the constituents in blood of normal and pathological cattle in 1931 but subsequently his publications have been few. He retired in 1964.

With the organization of the Faculty of Medicine in 1950 a Department of Biochemistry was immediately set up with Marvin Darrach as professor.

Marvin Darrach was born in Vancouver in 1913. His early education was at the University of British Columbia and he received a PHD from Toronto in 1940. He then joined the staff of Merck and Company, where he progressed to the position of director in the development of new products. He was appointed professor and head of the new Department of Biochemistry at the University of British Columbia in 1950. His research interests have been the chemotherapy of tuberculosis, steroid metabolism, and latterly hormonal control of antibody production. He retired in 1974.

The major contributions have been those of Gordon Henry Dixon, a native of South Africa, who did his doctoral thesis under C.S. Hanes in Toronto. He then worked with Hans Neurath at the University of Washington on the structure of trypsin and chymotrypsin as serine esterases (1954–58) and with H.L. Kornberg at Oxford (1959). He returned to Toronto, initially to the Connaught Laboratories, and then to the Department of Biochemistry where he began work on insulin and human haptoglobins. He joined the staff at the University of British Columbia in 1963. He has achieved an international reputation as a protein chemist for his work on the structure of haptoglobins, the partial synthesis of insulin, and the sequence of amino acids at the active centre of chymotrypsin. He was elected Fellow of the Royal Society of Canada in 1970. He became professor of biochemistry at the University of Sussex in England in 1972.

Sidney Howard Zbarsky joined the staff as associate professor in 1949 and has had full professorial status since 1962. He was born in Vonda, Saskatchewan, and obtained his BA in 1940 at the University of Saskatchewan, then his PHD in biochemistry at Toronto under Young in 1946. For two years he was employed at the Atomic Energy Project prior to his appointment to the staff at the University of British Columbia. He has prepared and made use of radioactive carbon compounds in studies of the metabolism of adenine and the biosynthesis of DNA in tumours. He has also investigated the activity of intestinal nucleases and the metabolism of BAL.

William James Polglase, a graduate of the University of British Columbia, was appointed to the staff in 1952 and made full professor in 1962. During the past twenty years he has published several papers, mainly on human glycogen, dihydroxystreptomycin and its activity as an antibotic. He is at present acting head of the department.

Gordon Malcolm Tener, a graduate of the University of British Columbia who had been associated with H.G. Khorana at the British Columbia Research Foundation (1954–60), joined the staff in 1960. He has accomplished meritorious research on antimycin A, sugar phosphates, and especially methods for the isolation, degradation, and synthesis of nucleotides.

MCMASTER

Biochemistry was taught in the Department of Chemistry from 1950 by Samuel Kirkwood, a Canadian from Alberta who had qualified for a PHD in biochemistry at Wisconsin. When he left Canada on appointment to the staff of the University of Minnesota in 1956, the teaching of biochemistry was taken over by I.D. Spenser.

Ian Daniel Spenser was born in Vienna in 1924. He attended the University of Birmingham where he qualified for a BSC in 1948 and London University (King's College) where he took a PHD in 1952. Then he lectured at St Bartholomew's Medical School from 1952 to 1957 before his appointment as associate professor in the Department of Chemistry at McMaster. Spenser's research has been in the biosynthesis of alkaloids and the metabolism of amino acids.

A separate department was formed in 1967 under R.H. Hall, who moved to McMaster that year. Ross Hume Hall is a Canadian from Winnipeg, where he was born in 1926. He attended the University of British Columbia (BA, 1948) and Toronto (MA, 1950). As an 1851 Exhibitioner he went to Cambridge University and qualified for a PHD in chemistry in 1951. On returning to Canada he was resident chemist at the Lederle Laboratories (1954-58), then principal scientist in cancer research at the Roswell Park Memorial Institute (1958-67) and associate professor of biochemistry at New York State University (1965-67). Hall then was appointed professor and chairman of the new Department of Biochemistry at McMaster. His major research interest is growth and differentiation, especially in regard to cellular mechanisms of processing information.

SUMMARY AND ANALYSIS

Departments of biochemistry were formed in 1968 at the Memorial University of Newfoundland and the University of Sherbrooke when faculties of medicine were organized there. The sequence of those who have been heads of departments of biochemistry or pathological chemistry at Canadian universities is shown in Table 2 from the dates of their formation until 1974.

The historical sequence of the teaching of biochemistry in Canada may be stated to have been in the following order: Macallum at Toronto from 1890 in the Department of Physiology, Ruttan at McGill from 1902 in the Department of Chemistry, Cameron at Manitoba from 1909 in the Department of Physiology, Harding at McGill from 1911 in the Department of Chemistry, Baril at Montreal from 1911 in the Department of Chemistry, Lothrop at Queen's from 1914 in the Department of Chemistry, Collip at Alberta from 1915 in the Department of Chemistry, Manning at Saskatchewan from 1916 in the Department of Chemistry, Young at Western from 1921 in the Department of Biochemistry, Bois at

TABLE 2
Universities and sequence of heads of departments of biochemistry and of pathological chemistry

University	Head of department	
BIOCHEMISTRY		
Alberta	1922-28	J.B. Collip
	1928-29	J.W. Scott
	1929-49	G. Hunter
	1949-61	H.B. Collier
	1961-	J.S. Colter
British Columbia	1950-74	M. Darrach
	1974-	W.J. Polglase (acting)
Dalhousie	1924-50	E.G. Young
	1950-65	J.A. McCarter
	1965-	C.W. Helleiner
Laval (Science)	1943-58	E. Bois
	1958-69	M. Jean
	1969-	P. Tailleur
(Medicine)	1942-63	J.R. Gingras
	1963-69	L. Berlinguet
	1969-	L.-M. Babineau
Manitoba	1923-47	A.T. Cameron
	1947-59	F.D. White
	1959-64	P.B. Hagen
	1964-74	M.C. Blanchaer
McGill	1922-28	A.B. Macallum
	1928-41	J.B. Collip
	1941-59	D.L. Thomson
	1959-69	K.A.C. Elliott
	1970-	A.F. Graham
McMaster	1968-	R.H. Hall
Memorial	1967-	L.A.W. Feltham
Montreal (Medicine)	1951-53	G.H. Baril
	1953-64	L.-P. Bouthillier (acting)
	1964-	W.G. Verly
Ottawa (Medicine)	1946-47	V. Vlassopoulos
	1947-48	D. Hingerty (acting)
	1948-49	M. Murnaghan (acting)

TABLE 2 (continued)

University	Head of department	
Ottawa (Medicine) cont'd	1949-61	J. Ettori
	1961-69	A. D'Iorio
	1969-	D.S. Layne
Queen's	1937-49	R.G. Sinclair
	1950-64	J.M.R. Beveridge
	1964-67	P.B. Hagen
	1967-	P.H. Jellinck
Saskatchewan	1946-49	H.B. Collier
	1949-67	C.S. McArthur
	1967-	J.D. Wood
Sherbrooke	1968-71	R.H. Despointes
	1971-	F. Lamy
Toronto	1907-17	A.B. Macallum
	1917-19	T.B. Robertson
	1919-29	A. Hunter
	1929-51	H. Wasteneys
	1951-60	A.M. Wynne
	1960-65	C.S. Hanes
	1965-70	G.E. Connell
	1970-	G.R. Williams
Western Ontario	1921-23	E.G. Young
	1924-47	A. Bruce Macallum
	1947-65	R.J. Rossiter
	1965-73	H.B. Stewart
	1973-	K.P. Strickland (acting)
PATHOLOGICAL CHEMISTRY		
Toronto	1910-15	J.B. Leathes
	1915-20	A. Hunter
	1920-35	V.J. Harding
	1935-47	A. Hunter
	1947-66	J.A. Dauphinee
	1966-	A.G. Gornall
Western Ontario	1951-60	E.M. Watson
	1960-72	A.H. Neufeld
	1972-74	J.C. Griffiths

TABLE 3
History and status of courses and degrees in biochemistry at Canadian universities*

University	Baccalaureate (with honours in)	BSC given since	MSC given since	PHD given since
Alberta	Biochemistry	1947	1932	1953
Bishops	Biochemistry	1968		
British Columbia	Biochemistry	1955	1951	1953
Brock	Biochemistry†	1968	1969	
Calgary	Biochemistry	1965	1965‡	1965‡
Dalhousie	Biochemistry	1958	1929	1955
Guelph			1965‡	1965‡
Laval	Biochemistry	1959	1940	1940
Manitoba	Biochemistry‡	1948	1926	1937
McGill	Biochemistry	1933	1923	1923
	Biochemistry & Chemistry	1962		
	Chemistry & Biology	1912-32		
McMaster	Biochemistry	1948	1962	1969
Memorial	Biochemistry	1967	1968	1970
Montreal	Biochemistry	1955	1947	1947
Ottawa	Biochemistry	1962	1961	1961
Queen's	Biochemistry	1940	1943	1954
Saskatchewan				
Regina	Biochemistry‡			
Saskatoon	Biochemistry	1950	1948	1956
Sherbrooke	Biochemistry (as major)	1965	1967	1968
Simon Fraser	Biochemistry§	1968	1965	1965
Toronto	Biochemistry	1919	1913	1903
	Biochemistry & Physiology	1910		
Victoria	Biochemistry & Bacteriology	1968	1970‡	1970‡
Waterloo	Chemistry	1962	1965‡	1965‡
Western Ontario	Biochemistry	1966	1925	1950
Windsor	Biochemistry	1958		
York	Chemistry & Biology	1965	1966‡	1966‡

* Courses in biochemistry are also offered at other universities or in other departments of the universities mentioned in the table: at Acadia, in Chemistry; at Alberta, in Field Crops; at Brandon, in Chemistry; at Dalhousie, in Biology; at Macdonald College, in Agricultural Chemistry and Animal Nutrition; at McGill, in Experimental Medicine and Investigative Medicine; at Mount Allison, in Chemistry; at New Brunswick, in Biology and Chemistry; at Ottawa, in Chemistry; at Queen's, in Chemistry; at St Francis Xavier, in Chemistry; at Concordia, in Chemistry; at Toronto, in Enzymology, Medical Research, Nutrition, and Pathological Chemistry; and at Western Ontario, in Pathological Chemistry.
† In Department of Biological Sciences.
‡ In Department of Chemistry.
§ In Department of Chemistry or of Biological Sciences.

TABLE 4
Origins of academic degrees held by biochemists cited in teaching and research in Canada (1907–70)

University	Baccalaureate*	Masters†	Doctorate‡
Acadia	6	2	
Alberta	16	12	
Brandon	2		
British Columbia	19	14	2
Carleton	1		
Dalhousie	8	12	
Laval	8		7
Loyola	1		
Manitoba	7	4	1
McGill	26	25	42
McMaster	2	1	
Montreal	9	4	5
Mount Allison	1		
New Brunswick	4		
Ontario Agricultural College	1		
Ottawa	2	1	
Queen's	6	2	
St Francis Xavier	1		
Saskatchewan	8	5	
Sir George Williams	1		
Toronto	51	34	63
Western Ontario	6	7	4
TOTAL	187	123	125
% OF TOTAL (250)	75	50	50

* BA, BSC, or BSA.
† MA, MSC, or MSA.
‡ PHD or DSC.

Laval from 1928. Other universities began to teach biochemistry much later, mostly with the establishment of independent departments of biochemistry. The sequence of establishment of such departments has been listed in Table 1. Thus the teaching of biochemistry originated from departments of chemistry or physiology and, in the newer universities, independently.

The sequence of the establishment of honour courses and advanced degrees in biochemistry is summarized in Table 3. In some cases the date is that when the first degree was conferred rather than when first offered. In several universities the honours course may be a combination of biochemistry with biology

(Waterloo, York) or chemistry (McGill, Simon Fraser). With the exceptions of McGill, Toronto, and Queen's, honours courses in biochemistry have been given mostly only since 1955.

It is significant of the growth of the discipline that an honours course is now offered in twenty universities in Canada. In contrast the master's degree in biochemistry was available much earlier (1923-32) in many universities, such as Dalhousie, Western Ontario, Manitoba, and Alberta, though the PHD was confined to McGill and Toronto. It only became more generally obtainable after 1950 and mostly after 1965. Departments were small and many universities hesitated to offer the advanced degree until recently.

This fact is reflected in an analysis of the academic qualifications of the 250 biochemists to whom reference has been made in this account. Table 4 shows the names of the universities in Canada conferring these degrees and the numbers of graduates. Approximately 75 per cent of all biochemists mentioned obtained their baccalaureate and only 50 per cent their post-graduate degree in Canada. Toronto and McGill, almost exclusively, have awarded the doctorate, although this situation is now changing. It is of interest that 35 (14%) had a qualifying medical degree (MB or MD). Of the 250 biochemists mentioned no less than 75 (30%) have been elected Fellows of the Royal Society of Canada and 12 (5%) Fellows of the Royal Society of London. It is also of interest to know those universities outside of Canada that awarded the doctorate to this group. There are thirty-four such institutions, including: London, which has awarded 15; Cambridge, 11; Wisconsin, 9; Minnesota, 7; Cornell, 7; Manchester, 5; Oxford, 4; Columbia, 4; Harvard, 6; Ohio State, 3; Edinburgh, 3; Liverpool, 3; Leipzig, 2; California, 2; Yale, 2; and many others which have awarded only 1. Thus, although the universities of London and Cambridge have attracted most postgraduate students, the United Kingdom and the United States of America have accounted for approximately equal numbers, 47 vs. 52. In contrast only 5 have obtained their degree in Germany and none in France.

PATHOLOGICAL CHEMISTRY

In only two medical schools in Canada (Toronto and Western Ontario) has the subject of pathological biochemistry been set up as a separate department.

Toronto
As early as 1910 such a department was instituted at the University of Toronto with John Beresford Leathes (1864-1956) as professor. Leathes (24-26) was a distinguished scientist who had been trained at Oxford and Guy's Hospital in London and who had taught physiology at St Thomas's Hospital Medical School,

associated with E.H. Starling, for nine years prior to accepting the new professorship at Toronto. He had also been associated with the Lister Institute and had done notable research on fat metabolism which was published in a classical monograph on *Fats* in the Longman series in 1910. He was elected a Fellow of the Royal Society of London and Canada in 1911. He gave two courses of lectures to students in the clinical years, one on general disorders of biochemical mechanisms and one on metabolic aspects of diseases of special organs such as pancreas, kidney, and thyroid. While in Toronto Leathes founded a local medical research society and established clinical biochemical laboratories at the Toronto General and St Michael's hospitals for the use of students in the chemical examination of material obtained from their cases on the wards. He also provided laboratory space in his department for the use of interns and graduate students for further training and research. This was located in the old Pathology Building, south of the Toronto General Hospital on University Avenue, but now demolished. He left Toronto in 1915 to become professor of physiology at Sheffield, where he remained until retirement in 1933.

Leathes was succeeded at Toronto by Andrew Hunter (q.v.), who held the chair from 1915 to 1920, during which period the department was called Chemical Pathology. Under his successor, V.J. Harding (q.v.), the departmental name was changed back to Pathological Chemistry. The department was moved to the Banting Institute in 1930, where it still is. Andrew Hunter again became professor in 1935 and continued as such until his retirement in 1947. J.A. Dauphinee succeeded him and he in turn on retirement was succeeded by A.G. Gornall in 1966.

James Arnold Dauphinee came from New Westminster, British Columbia, where he was born in 1903. He took degrees in the Faculty of Arts at the University of British Columbia before qualifying for the PHD in biochemistry at Toronto in 1929 under A. Hunter and the MD in 1930. From 1934 to 1941 he was a demonstrator in the Department of Medicine at Toronto. He served in the Canadian army (RCAMC) from 1941 to 1945 as a lieutenant-colonel, and was awarded the OBE in 1946. He was appointed an associate in medicine at Toronto in 1945 and professor of pathological chemistry in 1947.

Allan Godfrey Gornall was born in River Hebert, Nova Scotia, in 1914. He completed the honours course in chemistry cum laude at Mount Allison University in 1936 and received the PHD from Toronto in 1941 in the Department of Pathological Chemistry under Hunter. After serving as a research fellow in the department in 1941–42, he joined the services and acted as clinical biochemist to the Naval Medical Service in Halifax from 1942 to 1945, when he retired with the rank of lieutenant commander. Gornall returned to Toronto in 1946 to his old department as assistant professor and rose in rank to full professorship in

1963 and head of the department in 1966. In this capacity he initiated numerous changes in organization which restored its original clinical orientation and doubled its size and scope. The name of the department was changed to Clinical Biochemistry in 1972. Gornall's original papers concern the citrulline-ornithine cycle, intestinal perfusion in uraemia, surgical shock, liver function, glucagon, cortisone, and a lengthy study of the effects of aldosterone.

T. Frederick Nicholson joined the department in 1929 as lecturer and became full professor in 1962. He retired in 1967 and went to teach in Nigeria. He published a few papers on renal function as affected by kidney lesions.

Western Ontario
The clinical laboratories of the Victoria Hospital in London, Ontario, were organized in 1923 by Earle Macbeth Watson (1895-1973) as assistant professor of medicine and instructor of pathological chemistry at the University of Western Ontario. Watson was born in Belmont, Ontario, in 1895, and graduated in medicine from the University of Western Ontario in 1919. He did post-graduate work in both Edinburgh and London, England, prior to his appointment at Western, and also served as chemical pathologist at the Royal Infirmary and Jessop Hospital in Sheffield. He has published little of a biochemical nature. He instituted a course in 1934 which was in substance laboratory medicine with special emphasis on medical biochemistry. He was appointed professor of pathological chemistry in 1937 but a separate department was only created in 1951. The subject was taught in the laboratory of the North Building of the Victoria Hospital until transferred in 1972 to the new University Hospital on the main campus. Watson retired in 1960 and was succeeded by A.H. Neufeld.

Abram Herman Neufeld was born in Russia in 1907. From the University of Manitoba he received a BSC (1934), MSC (1935), and PHD in medical chemistry (1937). He was an instructor in biochemistry at Manitoba (1935-36), then lecturer at McGill (1936-41) and assistant professor of endocrinology there (1941-43). Neufeld obtained his MD from McGill in 1950. He was medical biochemist to the Queen Mary Veteran's Hospital in Montreal from 1946 to 1955 and chief of Biochemistry and Radio-isotopes service from 1955 to 1960. In 1960 he was appointed professor of pathological chemistry and chairman of the department at the University of Western Ontario. Neufeld acted as editor of the *Medical Services Journal of Canada* from 1947 to 1959, and as honorary secretary of the Canadian Federation of Biological Societies from 1961 to 1967. His original publications have been on the biochemistry of bromine, effects of trauma, the estimation of sodium and potassium by flame photometry, the proteins in myolomatosis, and adenosine triphosphatases. He retired in 1972 when a reorganization took place. The department became a division of clinical biochemistry with J.C. Griffiths, MB, BCH (Welsh National), in charge and it was separated from clinical pathology at the Victoria Hospital.

Richard Hugh Pearce, a graduate of Western, was appointed lecturer in 1950. From then until he resigned in 1961 to become associate professor in the Department of Pathology at the University of British Columbia he published numerous articles on the composition of human skin and connective tissues, especially with respect to their mucopolysaccharides, glycosamino glycans, hyaluronic acid esters, and hyaluronidase. With Watson he made a study of the increasing concentration of serum glycoproteins with age.

McGill

Mention should be made here of I.M. Rabinovitch (1890–) of the Montreal General Hospital who was also lecturer in biochemistry and toxicology at McGill. Between 1917 and 1947 he and his associates carried on numerous early investigations in the field of clinical biochemistry, especially metabolism in diabetes and its dietary control, tests of renal function, and a pioneering study of the metabolism of Eskimos in the Canadian Arctic. A Department of Metabolism at the old Montreal General Hospital was established for him in 1921, where he had A.F. Fowler and E.H. Bensley as associates. 'Rab' was largely responsible for the conception and development of the McGill-Montreal General Research Institute (q.v.) in 1945.

Biochemistry in agricultural colleges

Biochemistry was also taught in schools of agriculture in the early part of the twentieth century. Thus at Macdonald College a Department of Agricultural Chemistry was organized in 1907 with John Ferguson Snell (1870-1953) as professor. He was a Canadian, born in Snelgrove, Ontario, who had been trained as a chemist at Toronto and who took his PHD at Cornell in 1898. He worked for several years at the Storrs Agricultural Station (1898-1901) and assisted W.O. Atwater in constructing a bomb calorimeter for estimating the caloric value of foods. He also collaborated in compiling the well-known monograph on the chemical composition of American food materials (U.S. Dept. Agr. Bull. 28, 1899). His major interest in research at Macdonald was the chemical composition of maple products.

On his retirement in 1936 Snell was succeeded by William Douglas McFarlane (1900-75), a Glaswegian who received his university training in Canada. Courses at the Ontario Agricultural College, Guelph, earned him a BSA in 1925. Commuting between Guelph and Toronto, 'Scotty' managed to take the advanced courses offered by Andrew Hunter and H. Wasteneys and to be awarded the MA (1929) and PHD (1932). In 1930 he was appointed assistant professor in the Department of Biochemistry at the University of Alberta and in 1936 became head of the Department of Chemistry at Macdonald College. He was elected a Fellow of the Royal Society of Canada in 1952. He investigated several aspects of the composition of foods and nutrition, especially vitamins and some minerals. In 1939 he and G.H. Guest devised a colorimetric method for the estimation of proline which has been used extensively. McFarlane left in 1947 to take charge of the research laboratories of Canadian Breweries Ltd. in Toronto, where he stayed until retirement in 1965. He then undertook research for the Master Brewers Association of America at the University of California for a short time. He died at the age of seventy-four in Toronto in 1975. Robert Haddon Common, a graduate of Belfast and London, succeeded him at Macdonald. During the past twenty years various aspects of the metabolism of the domestic fowl have been investigated, with particular reference to the yolk proteins in laying and non-laying hens and the identification of avian hormones in urine.

On the subject of nutrition Earle W. Crampton has published numerous meritorious articles from Macdonald College. Crampton is an American by birth and education who came to Macdonald College in 1922, where he has since remained as professor of animal nutrition. His research work has been largely on the nutritive value of pasture herbage especially as related to the content of protein, the nutrition of the pig, the biological role of ascorbic acid, and the metabolic effects of processing of vegetable oils. With L.E. Lloyd he published a textbook entitled *Fundamentals of Nutrition* in 1959. Crampton was elected a Fellow of the Royal Society of Canada in 1945. On retirement in 1964 he was succeeded by Lewis Ewan Lloyd as head of a Department of Animal Science which amalgamated four others: animal husbandry, poultry, veterinary science, and animal nutrition. Lloyd resigned in 1967 to become head of the School of Home Economics in Winnipeg. He was succeeded by Herbert F. MacRae, a Maritimer who had been trained as a biochemist at McGill and who in 1973 became principal of the Nova Scotia Agricultural College in Truro, Nova Scotia.

At Alberta Robert Newton (1889-) was appointed Professor of Plant Biochemistry within the Department of Field Crops in 1922. He was primarily a cerealist and plant biochemist. He had graduated in agriculture from McGill in 1912 and had held several government appointments before obtaining the PHD degree in 1923 at Minnesota under R.A. Gortner. He resigned in 1932 to become the first director of the Division of Applied Biology and Agriculture at the National Research Council Laboratory in Ottawa, but returned to the University of Alberta in 1940 as dean of agriculture, to become later its president. No replacement was made on his departure in 1932, and plant biochemistry was taught to undergraduates from then until 1962 in the Department of Biochemistry by W.D. McFarlane (q.v.), and to graduates by A.G. McCalla in the Department of Field Crops until 1944.

In 1944 Arthur Gilbert McCalla became professor and head of the Department of Plant Science, in which department plant biochemistry is now taught. McCalla was born in St Catharines, Ontario, in 1906. He attended the University of Alberta where he obtained the MSC under R. Newton, and the University of California where he received the PHD under C.L.A. Schmidt. He has contributed mainly to the chemistry of the proteins of cereals, particularly the gluten of wheat.

At Manitoba, agricultural biochemistry has been taught by Allen Dinwoody Robinson (1904-71) since 1930 in the Department of Chemistry. He was born in Alliston, Ontario, in 1904 and received his education at the University of Saskatchewan and at Minnesota, where he took his PHD in 1930.

The teaching of biochemistry in the Faculty of Agriculture at the University of British Columbia has already been described in the section on that university. Other schools of agriculture have, of course, had courses of agricultural chemistry in the curriculum but have not contributed notably to new biochemical knowledge.

A.B. Macallum, first professor of biochemistry in Canada, at University of Toronto, 1908–17, and first chairman of the National Research Council, 1917–20 (see p. 5)

R.F. Ruttan, professor of chemistry, McGill University, 1902–27 (see p. 9)

A.T. Cameron, professor of biochemistry, University of Manitoba, 1920–47 (see p. 8)

A. Hunter, professor of pathological chemistry, University of Toronto, 1915–20 and 1935–47, and professor of biochemistry, 1919–29 (see p. 20)

V.J. Harding, associate professor of physiological chemistry, McGill University, 1917-20, and professor of pathological chemistry, University of Toronto, 1920-34 (see p. 10)

J.B. Collip, professor of biochemistry, University of Alberta, 1922-28, and McGill University, 1928-41 (see p. 16)

G.H. Baril, professeur de chimie physiologique, Université de Montréal, 1920-52 (see p. 29)

J.B. Leathes, first professor of pathological chemistry in Canada, at University of Toronto, 1910-15 (see p. 44)

E.W. McHenry, first professor of nutrition in Canada, at University of Toronto, 1927–61 (see p. 24)

J.H. Quastel, director of McGill-Montreal General Hospital Research Institute, 1949–66 (see p. 76)

R. Newton, first director of the Division of Biology, National Research Council, 1932–40 (see p. 49)

R.J. Rossiter, professor of biochemistry, University of Western Ontario, 1947– (see p. 27)

F.G. Banting, chairman of the Department of Medical Research, University of Toronto, 1923–41 (see p. 81)

E.J. King, professor of chemical pathology at the British Post-graduate Medical School, 1935–62 (see p. 98)

C.H. Best, chairman of the Department of Medical Research, University of Toronto, 1941–67, and professor of physiology, 1929–67 (see p. 81)

Medical Building, McGill University, where first course in clinical and physiological chemistry in Canada was taught in 1883 (see p. 5)

Medical Building, University of Toronto, where first department of biochemistry in Canada was located in 1908 (see p. 20)

Power House, University of Alberta, where biochemistry was first taught in 1915 by J.B. Collip (see p. 16)

Old Medical Building, University of Manitoba, where A.T. Cameron first taught biochemistry in 1909 (see p. 26)

Medical Sciences Building, Dalhousie University, which housed the Department of Biochemistry from 1924 to 1967 (see p. 27)

Old Medical Building, University of Western Ontario, which housed the Department of Biochemistry from 1921 to 1965 (see p. 26)

The old building of the Medical School of Laval University in the Quartier Latin in which biochemistry was taught from 1928 to 1957 (see p. 31)

Site of the original laboratory of Ayerst, McKenna and Harrison Ltd. on Craig Street in Montreal established in 1931 (see p. 51)

Interior of original laboratory of Ayerst, McKenna and Harrison Ltd. showing the first director, A. Stanley Cook, at left

Biochemistry in industry

In 1931 a research laboratory was organized at Ayerst, McKenna and Harrison Ltd. (now Ayerst Laboratories) in their original quarters in downtown Montreal by A. Stanley Cook (1906-67) with one assistant. It was concerned with bioassays of vitamins. Later research on hormones, sulpha drugs, and antibiotics was undertaken. Some of the earliest work on the preparation and properties of streptomycin was done there. Cook had been trained in chemistry at Dalhousie and had done post-graduate work under Harding at Toronto. He and Gordon A. Grant (1904-71) were responsible for developing what was probably the first industrial biochemical research laboratory in Canada. In 1974 it had a total professional staff of 147 (25 in the biochemical section) with research focused mainly on the discovery and examination of new drugs.

Roger Gaudry was assistant director, then director of research and vice-president of the company from 1954 to 1965, when he was appointed rector of the University of Montreal. He retired in 1975 both as rector and as chairman of the Science Council. Gaudry has had a distinguished career. He is a native of Quebec city and a graduate of Laval who won a Rhodes Scholarship (1937-39) and subsequently taught at Laval in the Department of Chemistry from 1940 till 1954. He was elected a Fellow of the Royal Society of Canada in 1954. Biochemical research at the Ayerst Laboratories is now under the direction of D.M. Dvornik, a Yugoslav by birth and education, who came to Canada in 1955 as a Post-doctorate Fellow of the National Research Council.

A control and research laboratory was organized at Chas. E. Frosst and Co. in Montreal in 1923 by Ezra Lozinski, who served as director until retirement in 1964. Lozinski was born in London, England, in 1896, but was educated at

McGill University, where he obtained the MD (1920) and the MSC (1923) in pharmacology. Under his direction the staff of the laboratory expanded from only one chemist to a total of sixty when he retired.

In 1914 the governors of the University of Toronto established an Antitoxin Laboratory in the Department of Hygiene under J.G. Fitzgerald. This became the Connaught Medical Research Laboratories in 1917 by reason of the gift of a 58-acre farm from Colonel A.E. Gooderham. The purpose of the laboratory was to prepare sera, vaccines, and other substances of importance in the practice of preventive medicine and to prosecute related medical research. It was thus comparable with the Pasteur Institute in Paris and the Lister Institute in London. From funds provided by the Rockefeller Foundation a building to house the laboratories and the School of Hygiene was built on the university campus in 1925. These laboratories have expanded enormously over the years and the various divisions are now in process of being consolidated at Willowdale, Ontario, where three new buildings were officially opened in 1966.

Notable fundamental investigations on insulin, penicillin, heparin and other antihaemostatic agents, carbonic anhydrase, desiccated fractions of human blood, and more recently a fibrinolytic proteinase from *Aspergillus oryzae* have been published by members of the staff of these laboratories, which have included Charles H. Best, David A. Scott, Albert M. Fisher, Arthur F. Charles, Mortimer D. Orr, Peter J. Moloney, and others. Methods for the commercial production of insulin have been under continuous investigation since 1922. The preparation and study of crystalline zinc insulin was achieved by David A. Scott and A.M. Fisher between 1935 and 1940. In 1955 Oliver Smithies developed the technique for separating proteins by electrophoresis in starch gels which has received international recognition. He is an Englishman, an Oxford graduate, who worked at the Connaught Laboratories between 1953 and 1960, when he began early fundamental studies of the genetic basis of variations in proteins. He left Toronto in 1963 to become professor of genetics at the University of Wisconsin. P.J. Moloney and M.D. Orr have made extensive studies of diphtheria toxin and toxoid over many years and also of the antigenicity of insulin. A.F. Charles has investigated the preparation and properties of heparin and other blood anticoagulants.

David Alymer Scott (1892-1971) deserves special mention because of his distinction as a biochemist (27, 28). He was born on a farm near Kincardine, Ontario, on 2 October 1892. In 1914 he entered the University of Toronto and specialized in chemistry and mineralogy. World War I caused him to work in an explosives plant and in the production of acetone with H.B. Speakman. In 1920 he graduated with an honours BA and in 1925 he obtained a PHD in biochemistry under Andrew Hunter for whom he always had a great admiration. In 1922 Scott joined the scientific staff of the Connaught Laboratories and for the following five years was engaged in the manufacture of insulin. In 1928, on sabbatical leave, he worked in the laboratory of Sir Charles Harington at the Institute for Medical Research in London in an effort to repeat the crystallization of insulin by J.J. Abel. In this they achieved success by using saponin. He also discovered Bertha Harington and they were married that year. Not until 1931 did Scott discover the role of zinc chloride in crystal formation and the reactivity of insulin with other heavy metal ions. With A.M. Fisher he developed the product known as Protamine Zinc Insulin. Later with A.F. Charles he worked out a method of producing the anticoagulant, heparin, from bovine lungs. He also contributed to the chemistry of carbonic anhydrase and of oxytocin. He retired in 1961. He was elected a Fellow of the Royal Society of Canada in 1939 and of the Royal Society of London in 1949. He was Flavelle medallist in 1954 and he was awarded the LLD (honoris causa) by the University of Toronto at a special convocation in celebration of the fiftieth anniversary of the discovery of insulin on 26 October 1971.

Unlike the chemical and pharmaceutical industries, the food industry in Canada has been very slow to employ professional biochemists and to do any research. It seems to have been content to depend upon laboratories in the federal departments of Agriculture and Fisheries or those of the National Research Council. All brewing companies and distilleries operate control laboratories and a few, such as Canadian Breweries and Labatts, carry on some research mainly on local problems. However, Canada Packers has maintained a research laboratory in Toronto which has recently been expanded.

Research at Canada Packers began in the early days of World War II when the supply of vitamin D from fish oils was inadequate and the company's feed business was endangered. William Flavelle McLean, a graduate in chemical engineering from the University of Toronto in 1937, started a search for other sources. This also took McLean to the presidency of the company in 1954. Synthesis of this vitamin from cholesterol, obtained from the brain and spinal cord of slaughtered animals, proved commercially successful and got research in the company off to a good start.

In 1952 the first building devoted entirely to research and development was completed and began operation with a staff of seventeen. In May 1966 a further extension made it one of the larger industrial research laboratories in Canada. There are twenty-five laboratories, four pilot plants, an experimental kitchen, large animal quarters, and sophisticated research equipment supplemented by a library of over 3000 volumes of books and journals. As of June 1973 the staff consisted of seventeen members with the doctorate degree, six with the master's degree, fifteen with the baccalaureate, and twenty-six technicians, for a total of sixty-four.

The director is Leon Julius Rubin, a graduate of the University of Toronto, who took a PHD under H.O.L. Fischer and Erich Baer in 1945. He joined the staff of Canada Packers that same year and worked on the synthesis of steroidal hormones and vitamin D. In 1949 he was appointed director of research. His personal research interests have included the synthesis and properties of pure triglycerides and glyceryl ethers.

The scope of the investigations at Canada Packers includes some organic syntheses, nutritional research, baking studies, investigations of meat curing, bacteriological studies of packaged meats and fish, studies on fats and oils, protein raw materials, and recently leather research. The studies of synthetic and semi-synthetic phospholipids as antithromboplastic agents which the laboratory has made and the development of stable, orally active anticoagulant and antilipaemic agents are of considerable interest and importance.

Biochemistry in government laboratories

NATIONAL RESEARCH COUNCIL

The National Research Council was established by Act of Parliament in 1916 with A.B. Macallum Sr. as the first chairman (1917-20) (29). It did not acquire adequate laboratories or staff until 1932, when a Division of Biology and Agriculture was organized with Robert Newton (q.v.) as director. When Newton returned to Alberta he was succeeded by W.H. Cook, who served as director until his retirement in 1968.

William Harrison Cook was born at Alnwick, Northumberland, England, in 1903 but was educated in North America. At Alberta he did research work with Newton and at Stanford with C.L. Alsberg. He joined the staff of the National Research Council in Ottawa in 1930 and remained with the council for thirty-eight years. He was interested throughout his scientific career in foods - composition, technology, and preservation - and for a period in weed killers, in association with many collaborators.

The name of the division was changed to Applied Biology in 1940, to Biosciences in 1964, and to Biology in 1968 when the divisions of Biosciences and Radiation Biology were combined with G.C. Butler (q.v.) as director and a biochemical laboratory set up under C.T. Bishop as assistant director. The number of the professional staff was 9 in 1932 and 112 in 1972 inclusive of Post-doctorate Fellows. Much of the research performed in the division has been biochemical in character and it has changed over the past thirty-five years from mostly applied projects to more fundamental research (30). It is again in transition.

The earlier work of the division was concerned with the processing and preservation of bacon, the malting qualities of barleys, the nature of rancidity, and the effect of hormones on wheat. With the advent of World War II problems of some urgency arose (31, 32), and interest turned to the preservation of human blood and of shell eggs, the refrigeration and dehydration of foods such as milk, meat, and eggs, baking tests for flour, the use of carrageenan from Irish moss as a substitute for Japanese agar, and a major investigation of the production of butanediol (2,3-butylene glycol) by fermentation of beet molasses with *Aerobacillus polymyxa*.

From 1946 to 1951 investigations of industrial fermentations continued and extended to waste sulphite liquor. Research on antioxidants in lard and shortening began. A prolonged study of the effects of a cold environment on animals was commenced, which later led to metabolic studies of Eskimos and the physiology of hibernation and hypothermia, by J.S. Hart and F. Depocas. The separation of wheat flour into starch and gluten was developed to the stage of a pilot plant by C.T. Bishop.

About 1952 the development of the technology of production of citric acid by fermentation of molasses was commenced, and also a series of investigations on chemical changes which accompany storage, freezing, and concentration of fluid milk. More fundamental work was undertaken in the section on macromolecules in studies of the hydrodynamic properties of the algal polysaccharides: alginic acid, laminaran, and carrageenan. The proteins of hen's egg yolk were separated and characterized as low and high density lipovitellins (α and β), phosvitin, and α, β, and γ livetins by ultracentrifugal, electrophoretic, and chromatographic methods. This investigation was under the direction of W.H. Cook assisted by D.B. Smith, W.G. Martin, R.W. Burley, G. Bernardi, F.J. Joubert, and others. The cellulases of *Myrothecium verrucaria* and the intermediate products therefrom in the hydrolysis of cellulose were investigated by D.R. Whitaker. In a long series of publications, G.A. Adams and C.T. Bishop studied the constitution of the hemicelluloses of wheat, oats, beechwood, white spruce, jack pine, birch, tamarack, and the polysaccharides of some bacteria and fungi.

In 1961 C.T. Bishop developed the first method for the gas-chromatography of sugars, D.R. Whitaker turned his attention to bacterial and fungal proteases and peptidases, and J.R. Colvin and S.T. Bayley began a fundamental study of the mechanism of formation of cellulose microfibrils in *Avena sativa* (oats) and *Acetobacter xylinum*. Methods of cultivation of unicellular algae in the laboratory were developed by P.R. Gorham in a study of strains of the blue-green, *Anabaena flos-aquae*, which forms a substance, saxitoxin, poisonous to animals and responsible for toxicity of edible molluscs at certain seasons of the year.

These investigations have been mostly accomplished by teams led by W.H. Cook in the work on yolk proteins, by R.W. Watson and D.R. Whitaker on enzymes, by N.E. Gibbons and G.A. Ledingham on fermentations, by G.A. Adams and C.T. Bishop on polysaccharides, by Dyson Rose and C.P. Lentz on the chemistry and technology of food processing, and by N.H. Grace and M. Kates on lipids.

The assistant director in charge of the Biochemical Laboratory is Claude Titus Bishop, mentioned above for his many fundamental contributions to carbohydrate chemistry. Bishop was born in Liverpool, Nova Scotia, in 1925. He obtained a BSC from Acadia University in 1945 and a PHD in organic chemistry from McGill University under C.B. Purves in 1949. He has been on the staff of the National Research Council ever since. He was elected a Fellow of the Royal Society of Canada in 1972.

The division has been located in the original building of the National Research Council on Sussex Boulevard since 1932.

In 1967 a Division of Radiation Biology was established within the National Research Council with Gordon C. Butler (q.v.) as director. It was fused with the Division of Biosciences in 1968.

The Prairie Regional Laboratory in Saskatoon was opened in 1948 and all investigators on the utilization of agricultural residues were transferred there from Ottawa. G.A. Ledingham (1903-62), a mycologist, was appointed the director in 1947 in succession to R.K. Larmour (1894-1970) who was first appointed. Over the intervening years this laboratory has been mainly concerned with the fundamental biochemistry and physiology of bacteria, mostly non-pathogens, and of many phytopathogenic fungi. This has been a very fruitful field of investigation led by A.C. Neish and R.H. Haskins, with F.J. Simpson and L.C. Vining as collaborators. From 1950 to 1955 the metabolic activities of *Ustilago zeae*, the causal fungus of corn smut, have been studied intensively. The ustilagic acids were isolated by R.U. Lemieux and characterized as a mixture of partially acetylated monoacidic β-cellobiolipids. They show some antibiotic properties. A search was made for moulds with high content of amylase, especially among the *Aspergilli*. The fermentation of glucose by *Bacillus subtilis* to produce glycerol was standardized. In 1953 R.U. Lemieux accomplished the synthesis of sucrose after an extended study of the effect of participation of the C2-substituent on the ease and stereochemical route of replacements at the anomeric centre of acetylated sugars, glycosyl halides, and glycosides.

The distribution of fatty acids in the oils of rapeseed, safflowerseed, etc., was determined by B.M. Craig, latterly with the aid of gas-chromatography. Precursors in the biosynthesis of lignin were established by the use of C^{14}-labelled compounds and also for quercetin, coumarin, and pungenin. The utilization of monosaccharides by *Aerobacter aerogenes* and *Leuconostoc mesenteroides* has been examined by F.J. Simpson (q.v.).

A study of the biosynthesis of the ergot alkaloids in *Claviceps purpurea* was initiated by L.C. Vining (q.v.) in 1959 and of the mustard oil glucosides by R.L. Wetter in 1964.

W.B. McConnell has studied the proteins of the wheat kernel, and E.M. von Rudloff has examined the terpenes and phenolics and waxes in various conifers, plant leaves, and woods.

The laboratory has thus been mainly concerned with the basic biochemistry of bacterial and fungal fermentations and the chemical constitution of many plant products: carbohydrates, proteins, lipids, terpenes, and polyphenols.

The Atlantic Regional Laboratory in Halifax was officially opened in 1952 with E.G. Young (q.v.) as director. He retired in 1962 and was succeeded by Arthur Charles Neish (1916-73). Neish was born in Granville Ferry, Nova Scotia, in 1916, a son of the rectory. He was educated at the Nova Scotia Agricultural College, Macdonald College, and McGill University, where he took his PHD in chemistry under H. Hibbert in 1942. He was appointed to the staff of the Division of Applied Biology of the National Research Council in Ottawa in 1943 and moved to Saskatoon when the Prairie Regional Laboratory was established in 1948. He became head of the section on plant physiology and biochemistry in 1956, and transferred to Halifax in 1961. With K.J. Freudenberg he published a two-volume treatise entitled *Constitution and Biosynthesis of Lignin* in 1968. His research work with many collaborators has included the metabolism of chloroplasts and of plant tumours, the production and properties of 2,3-butanediol by bacterial fermentations, the biosynthesis of lignin, cellulose, and xylan in the cell walls of plants, and a system for the analysis of the bacterial fermentation products of monosaccharides. In Halifax he became interested in the cultivation of the commercially important red seaweed *Chondrus crispus* (Irish moss) in the laboratory and then in a pilot plant which was built at Sandy Cove near Halifax. By using fortified sea-water and control of other parameters of growth the project was remarkably successful, and it has been adopted commercially.

Neish died in 1973, aged fifty-seven, after a rather prolonged illness with physical incapacity. During his life he received numerous honours including that of a special issue of *Phytochemistry* (vol. 12, August 1973) dedicated to him. He was elected a Fellow of the Royal Society of Canada in 1960 and a Fellow of the Royal Society of London in 1971. The National Research Council made him a Distinguished Research Scientist in 1973 on his retirement as director, and the Canadian government appointed him an officer of the Order of Canada the same year. 'Art' Neish was a shy, modest man and rather difficult to know well in consequence. He had a strong aversion to regimentation of research. Above all he was a dedicated scientist of great ability and penetration.

Frederick James Simpson was appointed to succeed Neish as director in 1973. Simpson was born in Regina and after receiving his initial education at the University of Alberta, took his PH D in bacteriology at Wisconsin. He joined the staff of the National Research Council in 1946 and did research on the bacterial metabolism of sugars and enzymic degradation of hemicelluloses at the Prairie Regional Laboratory from 1952 to 1970, when he moved to the Atlantic Regional Laboratory as assistant director.

One of the major biochemical projects at the Atlantic Regional Laboratory has been the investigation of various aspects of local seaweeds, a subject which had been little investigated in Canada previously. Most attention has been devoted to the polysaccharides: carrageenan, laminaran, agar, alginic acid, and furcellaran. Preliminary analyses of the approximate chemical composition of most of the more abundant species had been made by E.G. Young and M. Macpherson at Dalhousie University. Albert N. O'Neill (1920-58) worked out the approximate chemical structures of κ and λ carrageenans as polygalactan sulphates and synthesized laminaran sulphates. The former were shown to exhibit antipeptic and antithrombic properties and the latter to have antilipaemic and anticoagulant activities in blood by W.W. Hawkins and others. W. Yaphe discovered hydrolases for κ-carrageenan and agarose in some species of marine bacteria. D.G. Smith and E.G. Young determined the distribution of amino acids in the total proteins of representative species of algae and discovered free and combined ornithine and citrulline in *Chondrus crispus*. W.W. Hawkins investigated the relative nutritive values of seaweed meals. More recently attention has turned to the photosynthetic pathways in these plants and to some metabolites.

The only instance of the occurrence of pure chitin (poly-N-acetyl glucosamine) in nature was reported by J.L. McLachlan et al. in the flagella of a diatom, *Thalassiosira fluviatilis*, in 1964. More recently the biosynthesis of some antibiotics, such as sporidesmin, chetomin, gliotoxin, and chloramphenicol, has been the subject of research by L.C. Vining (q.v.) and A. Taylor.

ATOMIC ENERGY OF CANADA LTD.

In 1952 Atomic Energy of Canada Ltd., a crown company, was incorporated to take over from the National Research Council the plant and activities at Chalk River, Ontario, previously under its jurisdiction as the Atomic Energy Project with the ZEEP and NRX reactors. Dating from 1945 within this project there was a subdivision of biology and radiation hazards control under A.J. Cipriani (1908-56) which was engaged mainly in dosimetry and some effects of radiation in animals, including man. This subdivision had a biochemical section from 1948 onwards where radioactive materials were used as tracers in pioneering studies of several aspects of metabolism (G.C. Butler, J.A. McCarter, S.H. Zbarsky) and some radioactive compounds were synthesized (L. Siminovitch, F. Depocas). On the death of Cipriani in 1954 G.C. Butler was appointed director of the Division of Biology and Health Physics. Here many Canadian scientists were trained in the technology of radioactive materials.

Gordon Cecil Butler was born in Ingersoll, Ontario, in 1913. He was educated at the University of Toronto and took his PHD in 1938 with G.F. Marrian. Then as an 1851 Exhibition Scholar he worked at the University College Hospital in London with C.R. Harington between 1938 and 1940. On his return from England he was employed on the research staff of C.E. Frosst & Co. from 1940 to 1942, and then served in the Canadian Army as a major between 1942 and 1945. On demobilization he worked for a year at Chalk River in the Atomic Energy Project, and was then appointed to the staff of the Department of Biochemistry at the University of Toronto where he taught from 1947 to 1957. He then became director of the Division of Biology and Health Physics with Atomic Energy of Canada Ltd. In 1967 he was appointed director of the new Division of Radiation Biology of the National Research Council of Canada and in 1970 of the reorganized Division of Biology. He was elected a Fellow of the Royal Society of Canada in 1957.

Colin Ashley Mawson joined the staff in 1949 and is at present head of the branch on environmental research. For many years he has devoted his attention to the technology of disposal of radioactive wastes and to monitoring of areas of radioactivity. In 1965 he published a book on *Management of Radioactive Wastes.*

DEPARTMENT OF THE ENVIRONMENT

Fisheries Research Board
The Biological Board of Canada was constituted by Act of Parliament in 1912. It has served as a model in the later formation of other national research councils. The name was changed to its current title in 1937. The Board erected laboratories – which are known under various titles – at a number of locations: Nanaimo, British Columbia (1908); St Andrews, New Brunswick (1908); Halifax, Nova Scotia (1924); Vancouver (1942, previously at Prince Rupert from 1924); Grande-Rivière, Quebec (1936); Winnipeg (1944, moved to London, Ontario, in 1956 but closed in 1966 when a new Freshwater Institute was constructed on the campus of the University of Manitoba); St John's, Newfoundland (1948); St Anne de Bellevue, Quebec (1965); Burlington, Ontario (1967); Dartmouth, Nova Scotia (1968); and West Vancouver (1970). In 1970 the scientific staff numbered

297 and the total personnel 839. On 1 January 1973, the Fisheries Research Board became an advisory council free of administrative details and its research stations were integrated into the Fisheries and Marine Service of the Department of the Environment. The major biochemical investigations have been done at Halifax and Vancouver.

At Halifax the various methods of processing fish were studied initially: drying, salting, smoking, and freezing. The chemical nature of spoilage was later investigated by Stanley A. Beatty in his studies of the relation of trimethylamine and its oxide to bacterial enzymes. Beatty was director of the station from 1939 till his retirement in 1963. William J. Dyer and others have published numerous papers on the proteins of cod muscle, particularly in relation to rigor mortis and freezing. The technology of producing fish 'flour' (protein concentrate) was first developed in this laboratory. An extensive investigation of the fatty acids in fish and fish oils has been made by Robert George Ackman by gas-chromatography and refractometry. This may be considered an extension of the work done by H.N. Brocklesby (1901-63) (q.v.) at Nanaimo and Prince Rupert, British Columbia, between 1926 to 1938.

David Richard Idler was director of the station from 1963 to 1970. Born in Winnipeg in 1923, he graduated from the University of British Columbia with a BA in 1949 and an MA in 1950. He then took a PHD in biochemistry at Wisconsin in 1953. His appointment to the staff of the Pacific Fisheries Experimental Station followed, where he rose to the rank of director in 1961. In 1969 Idler was transferred to the station in Halifax and made the Atlantic Regional Director. When the Marine Sciences Laboratory was constructed outside St John's, Newfoundland, he was appointed to the staff of Memorial University as director of the new laboratory. He and his colleagues have investigated the steroidal hormones of molluscs and teleosts and their metabolism, especially in relation to migration of salmon. He isolated testosterone, free and combined, from the blood of the salmon and 11-ketotestosterone and 1-hydroxycorticosterone for the first time from any living form.

The present director of the Halifax station is E. Graham Bligh, who took his PHD at McGill in 1956 in biochemistry. After eight years of interrupted research work on extraction of fats at the Halifax Station (1951-53, 1956-62) he became associate editor of the *Journal of the Fisheries Research Board of Canada* and consultant to the chairman in Ottawa. Then in 1966 he moved to the Freshwater Institute in Winnipeg, where he studied the occurrence of mercury in fish. He returned to Halifax in 1971 as director.

The research station in Prince Rupert was established in 1925 to study the chemical and bacteriological changes occurring in fish during cold storage. A small staff, which included D.B. Finn, worked for a time in a primitive laboratory in a cold storage plant while a new building was under construction on a dock used by vessels of several fishing companies. On completion of the new station Finn was appointed acting director in 1927 and the staff was enlarged.

Donald Bartley Finn was born in London, England, in 1900. He attended the University of Manitoba, where he qualified for the BSC degree in 1924 and the MSC in 1928, and then Cambridge University, during a leave of absence from the station, obtaining his PHD in 1933. From 1934 to 1939 he was director of the Fisheries Research Station in Halifax, then successively chairman of the Salt Fish Board (1939-40), deputy minister of Fisheries (1940-46) and director of the Fisheries Division of the Food and Agriculture Organisation in Rome (1946-64). Since his retirement in 1964 he has been living in Italy. His contribution to biochemistry has been almost completely administrative.

When Finn took leave of absence in 1930 H.N. Brocklesby was appointed acting director, and in the fall of 1933 N.M. Carter (q.v.) became the director of the station, a post he retained until 1955. During World War II the facilities in Prince Rupert were taken over by the navy and the station moved to Vancouver in 1942, where it was initially located on Richards Street but later moved to the campus of the University of British Columbia.

Horace Nicholas Brocklesby (1901-63) (33) was born in Yorkshire, England, in 1901, and came to Canada with his family in 1911. He attended the technical high school in Winnipeg, and then worked as assistant chemist in the local plant of Lever Bros. Ltd. (1918-22) while taking courses in chemistry as a special student at the University of Manitoba, where he was awarded the BSC in 1926 and MSC in 1927. 'Brock,' as he was known to all, worked as research chemist at the Pacific Fisheries Experimental Station, Prince Rupert, British Columbia, from 1926 to 1938, during which period he acquired a PHD under Hibbert at McGill. When D.B. Finn, the director, was on leave of absence, 'Brock' served as acting director in 1930-32. He was chief chemist at the station from 1938 to 1942. With O.F. Denstedt he wrote a monograph of 150 pages entitled *The Industrial Chemistry of Fish Oils with Particular Reference to Those of British Columbia* which was published as Bulletin No. 37 by the Board in 1933. This was very well received and was followed by an enlarged edition entitled *The Chemistry and Technology of Marine Animal Oils with Particular Reference to Those of Canada* (Bulletin No. 59, 1944). He left to become co-ordinator and later director of research and development of the soy processing division of the Borden Co. In 1949 he established a private consulting practice in New York and in 1953 organized a company to produce concentrates of vitamins A, D, and E in Long Beach, California, for the feed industry. Brock became a Fellow of the Chemical Institute of Canada, a Fellow of the Royal Institute of Chemistry of Great Britain and Ireland, and a Fellow of the Royal Society of Canada. He was an authority on the chemistry and biochemistry of fats and oils and of the oil-soluble vitamins, and was instrumental in the formation of the Canadian branch of the American Oil Chemist's Society. After 1942 he became interested in protein supplements to animal feeds and to human dietaries. His hobby was music and while in Prince Rupert he conducted its philharmonic orchestra. 'Brock' died of leukaemia on 14 June 1963. Such a record describes a remarkably dynamic personality.

Neal Marshall Carter was born in Vancouver, in 1902, and obtained his BASC (1925) and MASC (1926) from the University of British Columbia. At McGill he took his PHD (1929) in organic chemistry under Hibbert; then he went to Germany for a year (1929-30) to Max Bergmann's Laboratory in the Kaiser Wilhelm Institut für Lederforschung in Dresden. There Carter developed a procedure for the synthesis of pure beta-monoglycerides. On returning to Canada, Carter was employed by the Fisheries Research Board at Nanaimo, British Columbia, where he and his associates made biochemical nutritional studies comparing the food values of different kinds of canned salmon and of other sea foods. In 1933 Carter became director of the Prince Rupert Station, and moved with it to Vancouver when the personnel and equipment were relocated there nine years later. In 1955 Carter was moved to the Board's headquarters in Ottawa as scientific assistant to the chairman and associate editor. He was succeeded by H.L.A. Tarr.

Hugh Lewis Aubrey Tarr was born in Clevedon, England, in 1905. He was educated at the University of British Columbia and took his doctorate under Hibbert at McGill in 1931. Then as an 1851 Exhibition Scholar he went to Cambridge, where he acquired another PHD in 1934. From 1934 to 1938 he was a research bacteriologist at the Rothamsted Experimental Station. He was then appointed to the staff of the Pacific Fisheries Experimental Station, where he progressed to the directorship in 1955. Tarr and his colleagues have been investigating various aspects of compounds containing phosphorus in fish and the enzymic systems concerned with their metabolism. Fish haemoglobins have also been studied. Tarr retired in 1970 and was succeeded by Neil Tomlinson as acting director.

Forest Products Laboratories
In 1913 the first forest products laboratory was established on the campus of McGill University in Montreal. Problems which arose in the use of Sitka spruce and Douglas fir in the aircraft industry during World War I led, in 1918, to the establishment of a second laboratory in Vancouver associated with the University of British Columbia. In 1927 the Montreal unit was moved to Ottawa and in 1958 new buildings were constructed there. Activities at the Vancouver and Ottawa laboratories are generally similar except that research on pulping is a Van-

couver responsibility. Large major chemical projects which have been undertaken include the preservation of wood against destructive agents, adhesion as in gluing and coating, the determination of chemical characteristics of Canadian woods, and pulping. Much of the work in these laboratories has consisted of routine testing or been applied in nature with little of a strict biochemical character. Until recently the staff has been small and uninspired. The most distinguished work has been that of Clara W. Fritz on the mycology of the pulp mill.

DEPARTMENT OF NATIONAL HEALTH AND WELFARE

Food and Drug Laboratories
The Department of National Health and Welfare is responsible for the enforcement of the Food and Drugs Act and Regulations first formulated in 1875 as 'an Act to Impose Licence Duties on Compounders of Spirits and to amend the Act Respecting Inland Revenue and to Prevent the Adulteration of Food, Drink, and Drugs' (34, 35). Local analysts were appointed in Halifax, Montreal, Toronto, and Quebec and later in St John, London, Winnipeg, and Ottawa, under the supervision of the Commissioner of Inland Revenue. Unfortunately there were no standards defining adulteration and the analysts did not possess the necessary knowledge. This led to the appointment of a chief dominion analyst at a salary of $2000 and the passing of the Adulteration Act in 1884, subsequently modified many times, often by Orders-in-Council. Henry Sugden Evans, PHC, FRMS, an eminent pharmacist from England, was thus appointed, but unfortunately he died two years later at the age of fifty-six. He was succeeded by Thomas Macfarlane, a Scotsman and a mining engineer previously attached to the Geological Survey as a field geologist. The annual appropriation for the whole department was $25,000 in 1886.

An examination for public analysts was instituted by an Act of Parliament in 1886 which also prohibited the appointment of anyone who did not possess a certificate of competency from the examining board. Among the first successful applicants were A. McGill and R.F. Ruttan in 1887. The examination lasted for ten days and only one candidate could be examined at a time. Anthony McGill (1847-1929) was then appointed as assistant analyst to Macfarlane, the dominion analyst, and thus began an association with the department which lasted for thirty-five years.

McGill was born in Rothesay, Scotland, in 1847. As a young man he came to Canada and attended the University of Toronto, from which he graduated with the BA in 1880 and BSC in 1882. He then taught science at the Ottawa Collegiate Institute for several years. When Macfarlane died suddenly in 1907, McGill succeeded him as chief dominion analyst. He was a very versatile man with strong research curiosity. From 1907 till his retirement in 1923 he established legal standards for foods and some drugs with the help of an advisory board via Orders-in-Council; he enlarged the staff with competent chemists and with district inspectors he assisted in the formulation of the new Food and Drug Act passed by Parliament in 1920; and he issued 440 bulletins of work done in the department. He was awarded the LLD (honoris causa) by the University of Ottawa in 1910 and elected a Fellow of the Royal Society of Canada in 1900. He died in Berkeley, California in 1929. He was unconventional in his habits of work, well read in English literature, and an amateur astronomer. He was an ardent chess player and fond of music. It is fair to consider McGill an eminent chemist primarily interested in biological material and thus one of the earliest biochemists in Canada.

Harry Mills Lancaster, BASC, succeeded McGill in 1923. He was trained as a chemist and came from Ontario. In 1919 the department came under the new Federal Department of Health and was named the Food and Drugs Division. A Laboratory of Hygiene was established in 1921 for research in public health and inspection of the manufacture and purity of some drugs with Norman MacLeod Harris, a professional bacteriologist, as the first chief. It was housed in the Annex of the Elgin Building with a small staff. The Food and Drugs Act of 1920 was amended in 1927 to impose stricter regulation of drugs. A section on pharmacology was instituted.

In 1945 Lancaster resigned because of poor health and Clarence Allison Morrell, a professional pharmacologist, succeeded him. In 1946 a directorate was organized to include the Food and Drugs Division, the Advertising and Labels Division, and the Proprietary or Patent Medicine Division with Morrell as director and chief dominion analyst.

The central laboratory for work on foods and drugs has been situated in various locations in Ottawa. The first was in the West Block of the Parliament Buildings (1884-1901) and the second in the Elgin Building, a two-storey structure converted from a hotel on Queen Street (1901-33). The equipment and personnel were moved to a much more commodious building, formerly a lumber mill, on John Street off Sussex Drive in 1933 which had previously been occupied by the National Research Council. Here a vitamin assay laboratory was built as an annex in 1937. In 1956 a new building was constructed as one of the first of the many to be built on Tunney's Pasture in Ottawa West and the first in the world designed and built for this special purpose.

The scope of research in the central laboratory in Ottawa has grown substantially from very modest beginnings to a healthy state with ninety-five on the professional staff in 1974 and major interests in the technology of vitamin assays (T.K. Murray), the measurement of the biological value of proteins (J.A. Campbell, A.B. Morrison), the metabolism of fatty acids found in rapeseed oil and marine oils, metabolic effects of non-nutritive sweetening agents (E.J. Middleton), medico-legal aspects of ethanol in breath (B.B. Caldwell), the detection of poisonous substances in foods (F.S. Thatcher, G.S. Wiberg), and the analysis of pharmaceuticals (M.G. Allmark, C. Farmilo). As noted above, the director of these laboratories has been known as the 'chief dominion analyst' since 1884 and appointees have been H.S. Evans (1884-86), T. Macfarlane (1886-1907), A. McGill (1907-22), H.M. Lancaster (1922-45), C.A. Morrell (1946-65) and R.A. Chapman (1965-71). The present incumbent is Alexander Ballie Morrison as assistant deputy minister in the Health Protection Branch. He had been previously engaged in nutritional research with the Food and Drug Directorate (1959-63). Successively he was appointed chief of the nutrition division (1963-66), chief of the pharmacological division (1966-68), director of research laboratories (1968-69), director-general (1969-71), and assistant deputy minister (1971-).

Two other members of the scientific staff of the Food and Drug Directorate merit special mention. Leonard Irving Pugsley retired as deputy director-general in 1970 after thirty-one years on the staff of the department. He was born in Five Islands, Nova Scotia, in 1900 and attended Acadia University and McGill University. From the latter he earned a PH D in 1932 in biochemistry under Collip. For two years he was a lecturer at Macdonald College (1934-36), and then assistant biochemist with the Fisheries Research Board at its Pacific Research Station (1936-39). His appointment as pharmacologist in the Laboratory of Hygiene of the Department of National Health followed in 1939 and he was promoted successively to chief of laboratory services for foods and drugs (1947-55) and then for the whole directorate (1955-58). Pugsley was made associate director in 1958 and deputy director-general in 1965. His scientific contributions have been very numerous and largely concerned with the assay of vitamins and hormones, especially vitamins A and D. With the Collip group at McGill he was a co-author of many papers on calcium and phosphorus metabolism and excretion as related to the parathyroid hormone. He was elected a Fellow of the Royal Society of Canada in 1962.

James Alexander Campbell retired in 1973 after thirty-two years of service with the Food and Drug Directorate. He obtained his PH D from McGill University in agricultural chemistry in 1947 under McFarlane. He was appointed as chemist in the vitamin laboratory of the directorate in 1941 and became its chief (1948-62). Thereafter he was successively director of research laboratories (1963-67), assistant director of foods (1967-71), and senior scientific advisor on foods (1971-73). With collaborators he has published numerous papers on vitamin assays, especially B_{12}, D, and niacin, and on the determination of the biological value of proteins in foods.

Division of Nutrition

As knowledge of human nutrition developed rapidly in the first quarter of the twentieth century and its national applications became apparent, the League of Nations encouraged the appointment of national committees to study and advise upon local conditions. Thus the Canadian Council on Nutrition was constituted in 1937 in the Department of Pensions and National Health with the deputy minister, R.E. Wodehouse MD, as chairman and C.A. Morrell, as secretary. The first meeting took place on 20 April 1938. Scientists on the first council were H.N. Brocklesby (FRB), H.F. Greenway (DBS), W.C. Hopper (Agr.), W.D. McFarlane (Macdonald), E.W. McHenry (Tor.), R. Newton (NRC), A. Stewart (Alta.), J.M. Swaine (Agr.), J.E. Sylvestre (Que.), F.F. Tisdall (Tor.), H. Wasteneys (Tor.), E.G. Young (Dal.), together with several consumers' representatives. With numerous changes in personnel this council met at least annually until 1969 and functioned as an advisory body within the department. It formulated a Canadian dietary standard in 1939 (36) which has been revised many times (37). The last extensive revision was in 1963 (38) but changes in the requirements of protein and iron have been made subsequently (39). The philosophy of dietary standards was extensively discussed over the years, which led to the original concept of a 'nutritional floor' below which normal good health could not be assumed and which entered into the Canadian formulations.

A dietary survey of families with low incomes in four cities was undertaken in 1939-40 in Halifax (E.G. Young), Quebec (J.E. Sylvestre and H. Nadeau), Toronto (E.W. McHenry and J. Patterson), and Edmonton (G. Hunter and L.B. Pett) as the first national effort in this field. A major conclusion of these surveys was that Canadian dietaries were deficient in thiamine based on a standard value which was later shown to be too high. This activity led to the formation of a permanent secretariat of the council in Ottawa in November 1941. Lionel Bradley Pett was appointed director of nutritional services and, in 1946, director of a new Division of Nutrition.

Pett was born in Winnipeg and educated at the University of Toronto, where he took a PHD in biochemistry with A.M. Wynne and later an MD while teaching at the University of Alberta. He served the council as secretary and later as chairman from 1941 until 1960. He organized numerous limited surveys of various kinds in British Columbia, Saskatchewan, Nova Scotia (Isle Madame and Sable Island), Quebec (Varennes and St Pierre), New Brunswick (Tracadie and Madawaska), Ontario (Timmins, South Porcupine, Elgin, and St Thomas), and Manitoba (St Vital) and one of the food habits of senior citizens between 1957 and 1963.

The contentious issue of the degree of milling of wheat and enrichment of flour which was desirable arose in 1940 when the council maintained a strong stand in advocating milling to an extraction of about 72 per cent without fortification and the adoption of vitamin B 'Canada Approved' flours and breads. This was in contrast to the practice of enrichment in the United States with B vitamins advocated by the Canadian millers for Canada, and that in the United Kingdom which required milling to 85 per cent extraction and fortification with iron and calcium carbonate to produce a gray, unpopular but nutritious bread. The vitamin B_1 (Canada Approved) flour contained a minimum of 400 IU of thiamine per pound of flour and produced a bread of 180 IU per pound.

In 1942 Canada's Food Rules for Health were formulated as a popular guide to adequate nutrition. With minor modifications these are still in use as Canada's Food Guide. The first edition of 'Table of Food Values Recommended for Use in Canada' was issued in 1943 and revised in 1951. A divisional laboratory was established in 1945-46 to provide facilities for the assessment of some parameters of health in nutritional surveys, especially concentrations of vitamins in food, blood, and urine, later extended as a diagnostic service to provincial departments of health and private physicians. This laboratory was transferred to Hygiene, now the Canadian Communicable Diseases Centre. The Canadian Bulletin on Nutrition was started in September 1948. The addition of potassium iodide to all commercial table salt was recommended for inclusion in the Food and Drug Act Regulations in 1944 and of ascorbic acid to apple juice and of vitamin D to fluid milk. A national survey of the weights, heights, and skinfold thicknesses of normal healthy Canadians was conducted between 1952 and 1957 by the division with the help of the Dominion Bureau of Statistics.

In 1960 J. Edward Monagle succeeded L.B. Pett as director of the Nutrition Division, and the division became more involved in the preparation of educational material and co-ordination of provincial activities in this field with a smaller staff as a result of recommendations of the Glassco Commission in 1962. The Canadian Council on Nutrition ceased to exist after 1969 but a pioneering national nutritional survey was undertaken by the directorate in 1969 after repeated recommendations of the council from 1964 onwards. The field work was carried on by two teams with a personnel of about twenty each and required two years, 1971-72. About 12,800 individuals were examined, with a disproportionate representation of pregnant women, Indians, and Eskimos. A preliminary report was published in November 1973 (40). No such comprehensive survey on a national scale had been undertaken in the world previously.

The national co-ordinator was Z.I. Sabry, and there were five regional directors. Sabry had previously been on the staff of the Department of Nutrition of the University of Toronto. He was born in Egypt in 1932 and educated at the University of Cairo (BSC, 1952), University of Massachusetts (MSC, 1954), and Pennsylvania State University (PHD, 1957). He taught at the American University of Beirut from 1957 to 1964 and was then appointed to the staff at the University of Toronto. He had had experience in a national survey in Lebanon in 1961.

Mention should be made here of the contributions of Fred F. Tisdall (1893-1950) (41) to Canadian nutrition. He was a paediatrician in Toronto associated with the Hospital for Sick Children from 1921 until his death. He was director of the research laboratory in the Department of Paediatrics from 1929 and published numerous articles with his colleagues, especially Allan Brown, Elizabeth Chant Robertson, H.D. Branion, J.H. Ebbs, T.G.N. Drake, and S.H. Jackson, on various aspects of nutrition. He was an original member of the Canadian Council of Nutrition and was reappointed for many years. He was instrumental in the organization of two surveys in Newfoundland in 1944 and 1948, the Canadian Red Cross Study of School Meals in 1947-49, and investigations of the nutriture of Indians in northern Manitoba and James Bay in 1942 and 1947 respectively. Tisdall was also involved with J.H. Ebbs and others in an important study of 400 expectant mothers in Toronto in 1941 which considered the relation of diet to their obstetrical histories. During World War II Tisdall joined the RCAF as squadron leader and took an active part in the administration of the food supply for the Air Force from 1940 to 1945. He set up four control laboratories at universities in Winnipeg, Edmonton, Guelph, and Montreal equipped to carry out food analyses. In 1945 they were turned over to the universities and have subsequently disappeared as nutritional laboratories. He was also the instigator of 'Canada Approved' flours and breads during this period. For one who was primarily a paediatrician these are remarkable accomplishments and bespeak an enthusiasm and energy which were characteristic of the man. He knew many people and worked well through political channels. His first investigation, done under John Howland at the Johns Hopkins Medical School with B. Kramer, immediately after graduation in 1916, resulted in analytical methods for calcium, magnesium, and phosphorus which were generally used for many years and channelled his interests toward calcium metabolism. Latterly he became an entrepreneur in the field of nutrition and lost some of the detachment required of the investigator.

DEPARTMENT OF AGRICULTURE

Agricultural Institutes
Much of agricultural chemistry may be considered biochemistry. There are agricultural colleges or faculties and experimental farms in all ten provinces. By the Experimental Farm Station Act a central experimental farm in Ottawa and four others were organized in 1886. In 1887 Frank B. Shutt was appointed as the only chemist at the Central Experimental Farm and he remained there until retirement in 1932 as dominion chemist in charge of the Division of Chemistry.

70 The development of biochemistry in Canada

This division became part of Science Service in 1938. In 1974 there were six research institutes, twenty-six research stations, and ten experimental farms, all under the jurisdiction of the Research Branch of the Federal Department of Agriculture. In 1974 the professional staff numbered over 910 and total personnel about 3850. Initially activities in research work were mostly of an applied character and were supervised by 'Science Service' under a director. Reorganization in 1964 set up eight institutes under separate directors and a director-general, Bert Baruch Migicovsky, who had been with the department since 1945.

Migicovsky made major contributions in bone metabolism, especially in the use of radioactive isotopes to follow movements of calcium and phosphorus into and out of the skeleton, and in the function of vitamin D in intestinal absorption. He was born in Winnipeg in 1915, and received his doctorate from Minnesota in 1940. He served in the RCAMC as major from 1943 to 1945. He then joined the staff of the Department of Agriculture at the Central Experimental Farm in Ottawa as agricultural scientist and has progressed through various positions to that of director-general since 1968. The assistant deputy minister for research, James Crawford Woodward, is also a biochemist in the field of animal nutrition who has been in administration in the department since 1947.

Much of the applied research work has been published in *Scientific Agriculture* (1921-45) or the subsequent *Canadian Journal of Scientific Agriculture* (1946-56) and since 1957 in the *Canadian Journal of Animal Science, Canadian Journal of Plant Science, Canadian Journal of Soil Science,* and *Canadian Journal of Microbiology.*

Grain Research Laboratory
In 1903 the Department of Agriculture established a cereal division at the Central Experimental Farm in Ottawa and in 1913 the Grain Research Laboratory in Winnipeg, under the Board of Grain Commissioners from 1926 onward. The laboratory has had the responsibility of grading the wheat produced under the Canada Grain Act of 1912 (revised in 1925) and reporting annually on its quality (42). F.J. Birchard (1875-1940) was appointed as chief chemist in 1913 and A.W. Alcock as assistant in 1914. Early studies were concerned with moisture in grain and its estimation, and the factors involved in the grading of wheat and flax. Because of internal administrative difficulties the laboratory was closed between 1923 and 1926. It reopened in the Grain Exchange Building in 1927 after clarification of its direction and legitimate projects, and the formation of the Associate Committee on Grain Research of the National Research Council. The testing of the protein content of grains became of great importance in blending and baking and the laboratory was equipped to make several hundred Kjeldahl determinations daily.

In 1933 Birchard retired. He was succeeded as chief chemist by W.F. Geddes (1896-1961), who was trained as a biochemist and had worked under R.A. Gortner and C.H. Bailey at Minnesota. The technologies of milling wheat and processing macaroni were investigated in relation to composition, and somewhat later micro-milling and baking techniques were studied. The first experimental work on the preparation and use of dry gluten to increase protein content and baking strength of flour was done in Winnipeg in 1937. In 1938 Geddes left to become professor of agricultural chemistry at Minnesota and later had a distinguished career in cereal chemistry.

John Ansel Anderson was appointed director in 1939. He was born in Sidcup, Kent, England, in 1903, but emigrated at the age of eighteen and was educated at the University of Alberta under Robert Newton and at Leeds University under W.H. Perkin Jr. He was on the staff of the National Research Council in Ottawa from 1931 to 1939. He then became chief chemist of the Grain Research Laboratory and held this position until he was appointed director-general of the Research Branch in the Department of Agriculture in Ottawa in 1963, where he remained until retirement in 1969. In Ottawa he was concerned with the technology of malting barley and published many papers on this subject (43). His studies on barley led to the development of apparatus and methods that placed the evaluation of malting quality on a sound basis. His discovery of the effects of lipoxidase in destroying the xanthophylls in durum wheat proved to be a major advance in testing such material. His work with bread wheats centred mainly on the rheological properties of bread dough and the effects of bromate and other factors thereon. His main collaborators were H.R. Sallans, W.O.S. Meredith, C.A. Ayre, I. Hlynka, and G.N. Irvine.

From 1939 onwards the principal grains studied were bread wheats, macaroni wheats, and malting barleys, mostly in relation to milling, baking, and malting. This has required investigation of the enzymes involved, mainly amylases, of oxidation on doughs, of carbohydrate-protein reactions, and of analytical methods. Such investigations are still in progress under the direction of George Norman Irvine as chief chemist since 1963, with Isidore Hlynka as assistant director.

Animal Research Institute
In the biochemical section under F. Sauer most research is basic in character and concerned with aspects of intermediary metabolism in domestic animals. Such investigations include the metabolism of selenium, the isolation of an inhibitor of the biosynthesis of cholesterol in hepatic mitochondria of rats, the pathways of synthesis of fatty acids, the structure of biological membranes, and the metabolism of carotenoids as affected by insecticides and herbicides. The present director is R.S. Gowe.

Mention should be made here of Arthur Raymond Gordon Emslie (1906-65) who was director of the Animal Institute from 1959 to 1964. He was born in Chefou, China, and was a graduate of the University of Toronto (BSA and MA in biochemistry under Wasteneys). He obtained the degree of DSC in 1934 from Aberdeen, where he worked under Sir John Boyd Orr at the Rowatt Institute on nutrition of poultry. On his return to Canada Emslie joined the then Science Service in the Federal Department of Agriculture and became chief of the Chemical Division in 1955. He died tragically in an automobile accident near Ottawa in 1965. His original scientific contributions were few.

Franz Aime Vandenheuval joined the staff of the Animal Research Institute in 1960 as a research officer. He had previously been employed by the Fisheries Research Board at its Halifax Station from 1947 where he had come from Belgium after World War II. Trained as a chemist in Brussels and London, where he had acquired a PHD in 1938 at Imperial College, he commenced work on marine lipids by analysis with a molecular still. This field he extended to blood lipids and lipoproteins particularly in relation to the structure and function of cellular membranes as in the myelin sheath of nerves. He has made major contributions to the technology of hydrogenation of fatty acids, to their partition chromatography and differential refractometry, and more recently to techniques in steroid chemistry, particularly their separation and identification on thin-layer chromatographic plates and on non-polar columns.

Food Research Institute
Projects of a biochemical nature at this institute include studies of the Maillard browning reaction and the detection of gibberellin-like compounds in apples, the physiology of lactic-acid bacteria associated with cheese-making, and the lipids of oil seeds and of the endosperm of wheat. R.P.S. Simms was the director from 1964 to 1973, when he became the research co-ordinator for research in field and oil-seed crops. The director is now M. Sahasrabudhe.

Chemistry and Biology Research Institute (previously Cell Biology)
Projects of biochemical interest are mainly related to the metabolism of phytopathogenic bacteria and fungi. Two organisms have been studied: *Agrobacterium tumefaciens*, the causative agent of crown-gall tumours of plants, and *Xanthomonas phaeseoli*, the aetiological agent of the common blight of beans. Glycolytic pathways have received special attention under the direction of R.M. Hochster (1922–71). α-5-Ribosyluracil has been isolated as a biosynthetic product in cultures of *A. tumefaciens*. Experiments have been carried out to distinguish differences in metabolism between tumorogenic and non-tumorogenic strains. A new

antibiotic, myxin, has been isolated from *Sorangium* which exhibits an unusually broad spectrum of inhibition. The director was R.M. Hochster until his untimely death on 16 September 1971. He was succeeded by G. Fleishmann.

Rolf Martin Hochster (1922-71) was born in Germany and completed his early education there in 1939, when he emigrated to Canada via England. He was employed by Chas. E. Frosst and Co. in Montreal from 1942 to 1947 and succeeded in obtaining the BSC degree with honours in chemistry from Sir George Williams University at the same time. In 1947-50 he was employed as a research fellow at the McGill-Montreal General Hospital Research Institute and received his PHD in biochemistry from McGill in 1950 under Quastel. From 1951 to 1956 he was an associate research officer with the National Research Council in Ottawa in the Division of Biosciences. He then became chief of the biochemical section in the Microbiology Research Institute of the Federal Department of Agriculture. In 1965 he became director of the Microbiology Research Institute and in 1967 director of the Cell Biology Research Institute. When the new Chemistry and Biology Research Institute was formed in 1971 he was named director of it. His publications have been concerned with enzymology of carbohydrates, chemistry of nucleic acids, metabolism of amino acids, biological transport systems, and latterly with phytopathogenic bacteria. In 1963 R.M. Hochster and J.H. Quastel edited a comprehensive treatise in two volumes on *Metabolic Inhibitors*.

Plant Research Institute
Projects of biochemical significance are numerous and diversified but mainly related to the tobacco plant, *Nicotiana tabacum*, and to wheat. R.B. Pringle and colleagues have published a series of papers on the chemistry and metabolism of toxins from plant pathogens, especially *Helminthosporium victoriae* and *Periconia circinata*. Other investigations include comparison of the composition of different species of *Fusarium* and of *Fomes* as wood-rotting fungi, and the formation of pigments by *Phleobopus* which are antibiotic. The director is Allen P. Chan.

Agricultural Research Institute (London, Ontario)
This institute was opened in 1950 with Hubert Martin as first director. Martin, an Englishman born in Kent in 1899 and a graduate of London University, was reader in biochemistry at the University of Bristol (1933-50) prior to his appointment as director of the new institute. He was succeeded by Elvins Yuill Spencer in 1960. Research interests have been concerned primarily with basic investigations of the mode of action of toxicants on plants and animals.

74 The development of biochemistry in Canada

Other Research Institutes
Investigations at the Soil Research Institute, the Entomology Research Institute, and the Research Institute at Belleville, Ontario, have not been primarily of a biochemical character. Federal agricultural research stations, concerned with local problems, have been set up in all provinces.

PROVINCIAL RESEARCH LABORATORIES

There are now provincial research councils in Alberta (1919), Ontario (1928), British Columbia (1944), Saskatchewan (1947), Nova Scotia (1946), and New Brunswick (1962), with their own laboratories (44). A Manitoba Research Council was established by legislation in 1963, and members appointed in 1964. As no laboratories are contemplated, the council serves to provide technical information, advice, and licensing. In Quebec a body designated Le Centre de recherche industrielle du Québec was constituted by statute in December 1969 and at the same time another designated l'Institut national de la recherche scientifique. Neither organization is more than administrative at present. The Newfoundland Research Council Act was passed on 1 April 1961, but no action has been taken to implement it as yet. Only in Ontario, Alberta, and British Columbia has there been any fundamental research of a biochemical character.

Nathaniel Hew Grace (1902-61) (45), an organic chemist by training but a biochemist in subsequent research work, served as director of the Research Council of Alberta from 1951 until his death in 1961. 'Nat' was born in Allahabad, India, on 10 November 1902, the son of the Reverend A.H. Grace, a missionary in India at the time. He attended schools in California and Saskatchewan, and then took honours in chemistry at the University of Saskatchewan, acquiring the BSC degree in 1925 and an MSC in 1927 under T. Thorvaldson. Then at McGill he took a PHD in 1931 under Otto Maass in physical chemistry. On graduation he was appointed to the staff of the National Research Council in the Division of Chemistry, where he investigated a wide range of subjects for twenty years. He did pioneering research on plant hormones which led to his transfer to the Division of Applied Biology. The advent of World War II diverted his work to problems of shortages of starch, rubber, resins, and vegetable oils. By the time that he left Ottawa for Edmonton in 1951 he had published over eighty scientific papers.

As director of the Alberta Research Council, Grace saw the staff expand to 130 individuals in 1956 and the construction of a new building for them on the campus at the University of Alberta. He was awarded the MBE in 1946 for his wartime investigations and elected a Fellow of the Royal Society of Canada in 1948, serving as president of Section V in 1958-59. He combined enthusiasm, vigour, and versatility with a lively sense of humour and innate love of mankind.

The microbiological leaching of metallic sulphide ores - 'biological mining' - has been under investigation for some years, led by Paul Chandos Trussell, the director of the laboratories in Vancouver. From 1952 to 1959 Har Gabend Khorana, a native of India, published an extensive series of important articles on synthesis of nucleotide esters, pentose phosphates, and polynucleotides, and achieved the total synthesis of coenzyme A. In 1960 Khorana and some of his group moved to the Institute for Enzyme Research of the University of Wisconsin despite efforts to retain him in Canada. He was awarded the Nobel Prize in Medicine in 1968.

An interesting series of papers on fatty acids with branched chains synthesized by *Bacillus subtilis* has been published by Toshl Kaneda from the Alberta Research Council in Edmonton.

Biochemistry in special institutes

MCGILL-MONTREAL GENERAL HOSPITAL RESEARCH INSTITUTE

This institute was organized by I.M. Rabinovitch (q.v.) in 1945 for research into special projects and cell metabolism as a unit of the Montreal General Hospital in association with McGill University. It was housed in an old stone residence on University Street opposite to the Medical Building. J.H. Quastel became the associate director in 1947 and director in 1949. In 1955 administration was taken over by McGill and the name changed to the above. In 1964 there was a reorganization and a dichotomy into a Research Unit of the National Cancer Institute under P.G. Scholefield and a Unit of Cell Metabolism under J.H. Quastel. During these years Quastel established a school with an international reputation which resulted in over 300 published articles in the general field of the fundamental action of drugs, especially but not exclusively on the nervous system.

Juda Hirsch Quastel was born in Sheffield, England, in 1899 and educated at the Imperial College of Science in London. He was later a Fellow at Trinity College, Cambridge (1924-29), director of research at Cardiff City Mental Hospital (1929-41), with the soil metabolism unit of the Agricultural Research Council at Rothamsted (1941-47), and then at the McGill-Montreal General Hospital Research Institute (1947-65). On retirement from McGill he was appointed professor of neurochemistry at the University of British Columbia and the Unit of Cell Metabolism at McGill ceased to exist.

In his early investigations at Cambridge 'Q' devised the technique of using 'resting' (non-proliferating) bacterial cells in studying their metabolic activities. At Cardiff he became interested in mental disorders and the mechanisms of drug action, especially in relation to oxidation and narcosis. He elaborated the concept of active centres in enzymic catalysis. At Rothamsted he collaborated in the early work with 2:4-D as a selective herbicide and the metabolism in soils. He and D.M. Webley proposed alginate as a soil conditioner. At McGill he continued his work on the action of drugs on enzymic systems and became interested in biochemical aspects of cancer cells. In collaboration he has published three books: *Neurochemistry* (1955) with K.A.C. Elliott and I.H. Page, *Chemistry of Brain Metabolism in Health and Disease* (1961) with D.M.J. Quastel, and *Metabolic Inhibitors*, 2 vols (1963), with R.M. Hochster. He has also contributed many original articles to scientific journals. He was elected a Fellow of the Royal Society of London in 1940 and of Canada in 1953. In 1970 he became a Companion of the Order of Canada. Quastel is thus one of our more eminent biochemists. Over seventy-five students have obtained their doctorate or been trained in research by him during the past twenty years, including M. Brassard, J.S. Colter, M. Franklin, M.J. Fraser, and P.G. Scholefield.

Peter Gordon Scholefield, an Englishman and a graduate of the University of Wales and of McGill, was associated with the McGill-Montreal General Hospital Research Institute from 1949 till 1957 and the McGill unit of the National Cancer Institute as director from 1959 to 1969, when he became assistant executive director of the National Cancer Institute. He has investigated mainly the metabolism of fatty acids in relation to phosphorylations and the energy transformations in the metabolic activities of malignant cells.

ALLAN MEMORIAL INSTITUTE OF PSYCHIATRY

This institute was founded in 1944 and set up in the Allan Reisdence on Pine Avenue in Montreal in association with McGill University and the Royal Victoria Hospital. In 1953 Theodore Lionel Sourkes was appointed as senior research biochemist. He and his associates have made many contributions to the difficult biochemical field of abnormal mental states. The major subject of investigation has been the metabolism of biogenic amines and the roles of decarboxylase and monoamine oxidase therein. This has related such amines to the aetiology of Parkinsonian epilepsy. Sourkes published a monograph on *Biochemistry of Mental Disease* in 1962. He was elected a Fellow of the Royal Society of Canada in 1971.

In this institute from 1958 onwards Catherine F.C. MacPherson has studied the immunoglobulins in normal and pathological cerebrospinal fluid with the help of immunoelectrophoretic analysis. She has devised a method of estimation and found a characteristic globulin present.

MONTREAL NEUROLOGICAL INSTITUTE

Some investigations of a biochemical nature have been carried out at this institute. In 1944 K.A.C. Elliott (q.v.) was appointed as neurochemist and he and his associates have published numerous articles on the metabolism of nervous tissue. He was co-author with I.H. Page and J.H. Quastel in writing a treatise on neurochemistry in 1955.

THE BANTING INSTITUTE (University of Toronto)

The first Nobel Prize (in Medicine) awarded for Canadian scientific work was given in 1923 to F.G. Banting and J.J.R. Macleod for the discovery of insulin. Banting shared his portion of the prize with C.H. Best and Macleod his with J.B. Collip. It is appropriate that a brief account of the work be given here, although it has been related many times (46). Some accounts are not entirely correct and some are not readily available.

During the academic year 1920-21, F.G. Banting (q.v.), a young surgeon who had not yet established a sufficient practice in London, Ontario, to support himself and was not welcomed by the local medical practioners who were mostly graduates of Western, augmented his meagre income by demonstrating in anatomy and physiology in the Medical School of Western University (now the University of Western Ontario). He also assisted F.R. Miller in the investigation of the effects of cerebellar stimulation. On the evening of 30 October 1920, while working on a lecture for the following day about the pancreas and its functions, Banting conceived the idea which altered his life and made him famous. As early as 1884 C.L.X. Arnozan and L. Vaillard had shown that ligation of the pancreatic ducts causes degenerative destruction of the acinar cells without apparent damage to the cells of the islets of Langerhans, whose function was not known. J. von Mering and O. Minkowski in 1889 discovered that extirpation of the pancreas induced a condition indistinguishable in animals from human diabetes mellitus. Textbooks published as late as 1920 discussed whether the islets produced an internal secretion or had some detoxifying influence. Banting wondered whether, if the antidiabetic principle were produced in the islet cells, previous failures to obtain active extracts might have been due to destructive effects of the digestive enzymes also secreted by the pancreas. Banting was eager to test his idea that, by ligating the pancreatic ducts of dogs and waiting until the zymogenic cells had atrophied, it might be possible to extract an active principle from the islets without its immediate destruction. The idea was not an entirely new one but it came to the right person at the right time.

Banting confided his idea to Miller, who declined to supervise research in a field so remote from his own. Miller advised Banting to consult J.J.R. Macleod, professor of physiology at the University of Toronto, who was an international authority on carbohydrate metabolism. Macleod was not impressed by the qualifications of an unknown young surgeon for research in a field that had baffled experts. However, it was finally arranged that Banting could have the use of a laboratory in Macleod's department for eight weeks, while he would be away on a visit to Scotland. Ten dogs were made available and Banting began work without any remuneration on 16 May 1921. The following day a young assistant, C.H. Best (q.v.), who had just finished his final examinations in physiology and biochemistry, joined him on Macleod's recommendation to help in the determinations of blood sugar. Best, too, worked without any stipend as Banting had no funds for paying him. Later Banting had to sell his car and other possessions to purchase dogs and other essentials needed to continue the research.

By 30 July an active extract had been prepared from the degenerated pancreas of a dog. When given intravenously, it lowered the blood sugar of a dog (depancreatized) from 0.20 to 0.11 per cent within two hours. Banting and Best, working day and night by themselves during Macleod's absence, had established by September that neutral, acidic, saline, or ethanolic extracts of degenerated canine pancreas consistently lowered the concentration of sugar in the blood of experimental diabetic dogs. These findings were reported to the Journal Club of the Department of Physiology at a regular Monday meeting on 14 November 1921. The results were first published in the *Journal of Laboratory and Clinical Medicine* (7: 251-66 [1922]). Banting and Best had been anxious to publish their first paper in the *Journal of the Canadian Medical Association* and were disappointed when Macleod insisted that it go to an American Journal of which he was an editor.

On 16 November 1921, a new era was begun when Banting and Best found that foetal calf pancreas did not contain pancreatic digestive juices but did contain insulin and that extracts made from them did not cause anaphylactic reactions when injected into dogs. Another milestone was reached on 12 December when an extract prepared from the pancreas of an adult cow, given intravenously, caused a drop in blood sugar from 0.28 to 0.11 per cent in three hours. The observation that ordinary bovine pancreas gave an active extract assured an adequate source of supply, if its effectiveness should be established in diabetic patients. These new results were also published in the *Journal of Laboratory and Clinical Medicine* (7: 464-72 [1922]). Meanwhile, difficulties had arisen in the preparation of a purified extract in quantities suitable for clinical trial. In December 1921, J.B. Collip (q.v.) undertook to investigate this biochemical problem. He was soon able to find a method of purification, based on fractionation with ethanol as described below.

To a small volume of 95% alcohol freshly minced pancreas was added in equal amount. The mixture was allowed to stand for a few hours with occasional shaking. It was then strained through cheese-cloth and the liquid portion at once filtered. The filtrate was treated with two volumes of 95% alcohol. It was found by this treatment that the major portion of the protein was removed while the active principle remained in the alcoholic solution.

After several hours the mixture was filtered and the filtrate concentrated to small bulk by distillation *in vacuo* at 18-30°C. The lipoid substances were then removed by twice extracting with ether in a separating funnel and the watery solution was returned to the vacuum still where it was further concentrated till it was a pasty consistency. 80% alcohol was then added and the mixture centrifuged. Four distinct layers were manifested in the tube. The uppermost was perfectly clear and consisted of alcohol holding all the active principle in solution ... The alcohol layer was removed by means of a pipette and was at once delivered into several volumes of 95% alcohol, or better absolute alcohol. It was found that this final treatment with alcohol of high grade caused the precipitation of the active principle along with adherent substances. Some hours after this final precipitation the precipitate was caught on a Buchner funnel, dissolved in distilled water and then concentrated to the desired degree by use of a vacuum still. It was then passed through a Berkfeld filter, sterility tests were made, and the final product was delivered to the clinic. (J.B. Collip, *J. Biol. Chem.* 55: XI [1923])

Extraction of bovine pancreas by this method was undertaken by the Connaught Laboratories. Under the guidance of Macleod a team was organized and the scope of the research was greatly enlarged.

On 11 January 1922, at the Toronto General Hospital, the first diabetic patient was treated with an extract of beef pancreas prepared by Banting and Best and previously tested for toxicity on themselves. The first public announcement of the successful operation on a human diabetic was made to a meeting of the Toronto Academy of Medicine on the evening of 7 February 1922. Because of the common occurrence of diabetes in the human population the demand for treatment with the new preparation was enormous.

In 1923 the Legislature of the Province of Ontario recognized the achievement by an unprecedented piece of legislation. On 4 May third reading was given to a bill providing for an annual grant of $10,000 to the University of Toronto for the promotion of medical research. As a result of this special provincial grant, the Board of Governors of the University of Toronto established the Banting and Best Chair of Medical Research and appointed Banting to fill it, a post he held until his death eighteen years later. He was succeeded by his former colleague, C.H. Best, in 1941.

Although Frederick Grant Banting (1891-1941) cannot be classified as a biochemist he contributed significantly to biochemical knowledge and for such deserves mention here. He was born at Alliston, Ontario, on 14 November 1891. He entered Victoria College (University of Toronto) in 1911 but transferred to the Faculty of Medicine in 1912. In 1915 he enlisted as a private in the Canadian Army Medical Corps but was sent back to finish his medical studies, which he accomplished in 1916. Banting again enlisted and went overseas with No. 3 Canadian General Hospital in July 1918. At Cambrai on 28 September he was severely wounded and received the Military Cross for gallant conduct on the field.

After the war Banting acted as resident surgeon in the Hospital for Sick Children in Toronto for a year, and then commenced practice in London, Ontario, in 1920. The discovery of insulin has been described above. The first regular clinic for the treatment of diabetes with insulin began in May 1922, at the Christie Street Hospital in Toronto. Banting at that time chose a life of medical research rather than one of medical practice and accepted the chairmanship of the new Banting and Best Department of Medical Research in 1923. He received numerous honours thereafter, including Fellowship of the Royal Society of Canada in 1926 and of the Royal Society of London in 1935. In 1934 he was created a Knight Commander in the Civil Division of the Order of the British Empire.

At the age of thirty-two Banting gathered about him a group of promising young workers who were located in the Banting Institute when it was opened in 1930. His subsequent career and tragic death in February 1941 have been described elsewhere (47, 48).

Charles Herbert Best was born in West Pembrooke, Maine, in 1899, the son of a Canadian physician. He was educated at the University of Toronto in science (BA, 1921; MA, 1922) and in medicine (MD, 1925). He then spent several years in London at the Medical Research Council Laboratories in Hampstead with Sir Henry Dale and acquired the DSC (London) in 1928. He was appointed professor of physiology at Toronto in 1929 but also retained his association with the Banting Institute and was its director from 1941. Best has been the recipient of innumerable honours of various kinds from all over the world. He was elected a Fellow of the Royal Society of Canada in 1931 and of London in 1938. Although primarily a physiologist, professionally, his investigations have had much biochemical significance. Studies of the mode of action of insulin occupied his earlier years. This led to investigations of hepatic metabolism of carbohydrates and then of lipids in collaboration with C.C. Lucas. The role of choline and lecithin in fat metabolism was elucidated and the aetiology of 'fatty' livers. Choline was established as a dietary essential and other lipotropic factors investigated, such as biotin and inositol. Heparin was studied in its relation to blood clotting. Best has written an account of his work to 1959 (49). With N.B. Taylor he wrote two very successful textbooks, *The Living Body* (1938) and *Physiological Basis of Medical Practice* (1937), which have passed through many editions and translations.

Funds poured in from a variety of sources to help establish the new department in the University of Toronto. It is not generally known that the department was set up outside the Faculty of Medicine, to assure freedom from possible restricting regulations of that faculty at that time. Members of the new department were not only exempted but excluded from formal teaching responsibilities. Research under the aegis of the new chair began in July 1923, in a room provided by Velyen E. Henderson of the Department of Pharmacology in the Medical Building. The personnel consisted of Banting, Miss Sadie Gairns (his chief assistant and secretary), and a laboratory boy. B.S. Cornell joined them in studies of adrenal insufficiency and the following year G.H.W. Lucas was added to the team. In January 1926 the single room was abandoned for three rooms on the ground floor of the old Pathology Building on University Avenue. On 21 November 1928 work began on a six-storey building to house those departments of the medical faculty that needed to be close to the Toronto General Hospital. The fifth floor was reserved for the Department of Medical Research. At the suggestion of Sir William Mulock, chancellor of the university, the new building was named the Banting Institute. Each department was autonomous, and Banting was not the director of the institute which was formally opened on 16 September 1930.

The origins of the Banting and Best Department of Medical Research have been given and the location and nature of the Banting Institute have been described. On 22 July 1925, the Banting Research Foundation was incorporated to assist financially medical research at the University of Toronto or elsewhere, and to provide a constant source of funds for the Department of Medical Research.

BANTING AND BEST DEPARTMENT OF MEDICAL RESEARCH

Since 1923 many areas of study have been pursued in this department. The earliest investigations concerned adrenal insufficiency, pernicious anaemia, and cancer. About 1926 Banting organized a team to study silicosis, a serious problem in Ontario at the time, and different facets of this disease were examined over a period of more than ten years. Meanwhile studies of calcium and phosphorus metabolism, of phosphoric esters in normal and malignant tissues, and of phosphatases, cholinesterases, and other enzyme systems were made. Enzymes in the stools of infants with intestinal intoxication, lead poisoning in infants, glutathione and other thiols, chemoantigens, royal jelly, and chemically induced shock-therapy all received study. Two new medical tools, electroencephalography and chemotherapy (with prontosil, sulphanilamide, and other sulpha drugs), received the first Canadian study in the department. In 1938, with the threat of war predominating, studies of the effects of simulated high altitudes and protection against agents of chemical warfare began. About 150 new sulpha drugs were synthesized in a search for a cure for tuberculosis. The same groups worked together later to develop the production of penicillin from the test-tube to the bottle stage and finally to the pilot-plant stage. This experience was transferred, with some of the key personnel, to the Connaught Laboratories, where the large-scale production of penicillin was soon underway. The distribution of arsenic in the tissues of animals exposed to lewisite was determined.

Banting, as chairman of the Associate Committee of Aviation Medical Research of the National Research Council, had been primarily concerned with problems of airmen. Best, as director of the Medical Research Division of the Royal Canadian Navy, brought new kinds of problems to the department, such as motion sickness, night vision, and biochemical changes after burns and in shock. Later Best directed or co-ordinated studies on fat absorption and fat metabolism, fatty livers and other effects of choline deficiency, other lipotropic phenomena, effects of diet on cholesterol in blood, liver, and other tissues, experimental diabetes, tissue changes induced by growth hormone and glucagon, effects of oral hypoglycaemic agents, and on numerous other projects. In this department, Erich Baer (q.v.) and his colleagues have made available to biochemists a large number of pure, optically active phospholipids and phosphonolipids and related compounds.

In 1937 H.O.L. Fischer (1888-1960), eldest son of Emil Fischer, along with Erich Baer, his former student, emigrated from Germany to Canada because of the rising Nazi power. They became members of the staff of the University of Toronto associated with the Banting Institute after a personal invitation from Banting. Fischer (50, 51) was born in 1888 in Würzburg, Bavaria, and educated in Cambridge and Berlin. He received his PHD in 1912 under Ludwig Knorr at the University of Jena, and after World War I taught chemistry at Berlin (1918-32) and Basle (1932-37). He carried on research in Toronto from 1937 to 1948, when he moved to the University of California at Berkeley, where he later became head of the Department of Biochemistry and remained active until his death in 1960. Although primarily an organic chemist much of his work pertained to substances of biochemical importance such as depsides, asymmetric glycerides, inositols, and simple monosaccharides. During his period in Toronto syntheses of asymmetric mono-, di-, and tri-glycerides, glycerophosphates, and glyceryl ethers were accomplished.

Erich Baer remained in Toronto and has further developed the synthesis of optically active derivatives of glycerol which has led to the synthesis of many phosphatides, cerebrosides, phosphatidyl glycols and peptides, and phosphonolipids, for which he has received several honorary awards. Baer was born in Berlin in 1901 and received his PHD from the University of Berlin in 1927. He was associated with H.O.L. Fischer for many years and accompanied him to Toronto in 1937. One of their outstanding accomplishments was the synthesis of *dl*-glyceraldehyde-3-phosphate, now known as the Fischer-Baer ester, which has helped to establish the Embden-Meyerhof scheme for anaerobic glycolysis to lactic acid in muscle. He has thus moved from simple carbohydrates to complex lipids with glycerol as the key to the transition. He was elected to Fellowship of the Royal Society of Canada in 1956 and was Flavelle medallist in 1966.

Colin Cameron Lucas was on the staff of the Banting and Best Department of Medical Research from 1934 until retirement in 1969. He was born in Winnipeg in 1903, and studied at the University of British Columbia, from which he was awarded the BASC in 1925 and MASC in 1926. During this time he spent five summers at the Nanaimo Station of the Fisheries Research Board doing analyses of sea-water. He taught chemistry for one year at Brandon College before commencing studies for his doctorate at the University of Toronto in 1927. He was the first student in Banting's department and was thus connected with the department for a total of forty-two years. For the reasons given above Lucas was technically registered in the Department of Pathological Chemistry under V.J. Harding, and shared a bench with E.J. King (q.v.) in that department. His research work, on some sulphur compounds of biochemical importance, directed by George Hunter, did not progress satisfactorily. Lucas, however, completed the course requirements for the PHD degree. For five years (1929–34) he again taught at Brandon College during the winter and did research on silicosis in Toronto in Banting's laboratory during the summer. In 1934 he joined the staff permanently and was promoted regularly to full professorship in 1946. After he received the PHD degree in 1936 he spent one year in London with C.R. Harington at the Institute for Medical Research. There followed many years of investigation in Toronto, often in collaboration, on such topics as royal jelly, metabolism of sulphanilamide and many other sulpha drugs, the production of penicillin, the composition of renal calculi, and most extensively on lipotropic phenomena including the influence of dietary protein on the metabolism of ethanol.

CHARLES H. BEST INSTITUTE

By 1950, congestion in the Banting and Best Department of Medical Research and in the old Medical Building at the University of Toronto was severe. Gifts from Canadian, British, and American pharmaceutical firms, from government sources, and from numerous individuals concerned with training and research in medicine made possible the erection of a mate to the Banting Institute. The Charles H. Best Institute, a four-storeyed building, was constructed on College Street, adjacent to the Banting Institute, and was opened in September 1953 by Sir Henry Dale.

The new building housed most of the 'B & B' Department of Medical Research and the Department of Physiology. The main fields of interest continued to be experimental diabetes, fat metabolism, lipotropic phenomena, and blood clotting. Effects of exposure to cold and radiation were examined and numerous other studies more physiological than biochemical have been conducted by its staff. The biochemical, histochemical, and nutritional studies that concern this history were carried out by Colin C. Lucas, Jesse H. Ridout, Jean M. Patterson, Robert J. Young, K.K. Govind Menon, R.J.G. Gillespie, Sailen S. Mookerjea, W.G. Bruce Casselman, James Campbell, James M. Salter, Bruno Rosenfield, Jessie M. Lang, Arnis Kuksis, Cecil C. Yip, and their associates.

In 1967 Best retired after twenty-six years at the helm of the department. The co-discoverers of insulin between them chaired the department for forty-four years, during which period over 870 papers were published from the department, on which appeared the names of over 220 investigators. The present director is Irving B. Fritz.

HOSPITAL FOR SICK CHILDREN RESEARCH INSTITUTE, TORONTO

In 1918 funds were donated by Mrs W.C. Teagle to establish a small chemical laboratory in the hospital then on College Street and built in 1875. It was equipped by a donation from Sir Joseph Flavelle with the directive that it should be engaged in chemical research. Miss A. Courtney, BA, was appointed the first director and early interest was in nutrition. In 1929 F.F. Tisdall (q.v.), a clinical assistant in the hospital, succeeded her and under his direction notable advances were made in the field of infant nutrition with the development of such foods as Pablum, Mead's Cereal, and Sunwheat Biscuits. Money from royalties on the Pablum patents became available about 1935 and this was used for the support of research projects and increased professional staff. Appointment of Elizabeth Chant Robertson and S.H. Jackson resulted.

During World War II the laboratory co-operated with the federal government to assay the vitamin content of wheat, which led to the introduction of 'Canada Approved' breads and flours. After the death of F.F. Tisdall in 1949 T.G.H. Drake was appointed to succeed him. The laboratory was moved in 1951 to the new hospital on University Avenue and a Research Institute was established as a separate department within the hospital. A.J. Rhodes succeeded Drake in 1953 as director. The Institute was enlarged in 1972 when the Elm Street wing was opened.

Biochemical research has played a prominent part in the investigations of the laboratory since 1929. The first biochemist, appointed in 1937, was S.H. Jackson, who took a PHD in pathological chemistry at the University of Toronto under V.J. Harding in 1936. He became director of biochemical research in 1950 and biochemist-in-chief in 1970 as well as professor in the Department of Clinical Biochemistry at the university and associate director of the institute. His early investigations were concerned with the B vitamins in the milling of wheat. More recently he has studied metabolism after thermal burns. He has devised a bilirubinometer and a single chilometric titration method for the determination of calcium and magnesium in blood serum. In 1973 his division in the institute included eight senior investigators and two research fellows. Current projects are concerned with collagen metabolism, genetic defects, epithelial and other glycoproteins, structure of myelin and membrane proteins, and regulation of secretion of insulin and growth hormone.

NATIONAL CANCER INSTITUTE

The National Cancer Institute of Canada was formed in 1947 with the financial support of the Canadian Cancer Society and the Department of National Health and Welfare. It makes grants-in-aid and appoints Fellows and Associates. It supports research units at the universities of Saskatchewan (1957-70), British Columbia (1960-), Alberta (1961-), Western Ontario (1961-), and McGill (1964-) and has assisted the Ontario Cancer Institute in Toronto and the Montreal Cancer Institute at Hospital Notre-Dame, since 1948. For many years it has made substantial grants to the McGill-Montreal General Hospital Research Institute under J.H. Quastel (q.v.) and the Manitoba Cancer Research Foundation under L.G. Israels.

Research at these units has contributed to biochemical knowledge. At the University of Saskatchewan Joseph Frances Morgan has made important studies of the specific nutrient requirements of fibroblasts in tissue culture. This has led to a series of publications on the antitumour activity of fatty acids. The technique shows promise as a tool for the investigation of the biochemical nature of malignancy.

Morgan was born in Vancouver in 1918 and received his early university training at the University of British Columbia, but his PHD at Toronto in biochemistry under A.M. Wynne. For one year he was employed as a biochemist in the Laboratory of Hygiene in Ottawa and from 1947 to 1952 as a research associate in the Connaught Laboratories in Toronto. He returned to the Laboratory of Hygiene as research biochemist in 1952 and was promoted to chief in 1959. In 1962 Morgan was appointed director of the Cancer Research Unit at the University of Saskatchewan and professor and head of the Department of Cancer Research at the university. In 1970 the unit was discontinued but the department has been maintained by the university. Morgan was elected a Fellow of the Royal Society of Canada in 1961.

At the University of Western Ontario and previously at Dalhousie J.A. McCarter (q.v.) has developed and extended earlier techniques for the study of chemical carcinogenesis in skin.

At the University of Alberta the Cancer Research Unit is located in the McEachern Laboratory in the Medical Sciences Building. The initial plan was agreed upon in 1959 and the unit was officially opened in 1963. Senior staff members hold academic appointments in the Department of Biochemistry and assist in teaching. A.R.P. Patterson has been the director since the unit was opened. He graduated with a PHD in biochemistry from the University of British Columbia in 1956 and taught there as associate professor from 1958 to 1962. Investigation is mainly related to various aspects of the synthesis and metabolism of nucleotides, nucleosides, and their analogues. The mechanism of action and of resistance to antitumour agents is of special interest.

Another active member of the staff is Joseph Franklin Henderson, who came from the United States and graduated with the PHD in oncology from the University of Wisconsin in 1959. He and his associates have published numerous studies on the metabolism of purines, the phosphoribosyl transferases, and inhibitors of biosynthesis in Ehrlich ascites tumour cells.

At the Ontario Cancer Institute in Toronto H.E. Johns and L. Siminovitch have made biophysical studies of the effect of ultraviolet rays on nucleic acids and their enzymes. Bernhard Cinader, head of the subdivision on immunochemistry and professor of cell biology at the university, has contributed to the immunochemistry and genetics of mammalian polymorphic proteins. He was born in Vienna in 1919 but educated at the University of London. From 1945 to 1953 he worked with R.A. Kekwick at the Lister Institute and emigrated to Canada in 1958. He was elected a Fellow of the Royal Society of Canada in 1971. His study of immunological tolerance led to his theory of a 'steering' mechanism which explains the relationship between the specificity of the antibody response and tolerance to autologous macromolecules. He discovered a defect in complement in inbred strains of mice which led him to define the genetic criteria for inborn errors of metabolism.

The Montreal Cancer Institute was formed in 1948 and directed until 1967 by Antonio Cantero and subsequently by Roger Daoust, a histologist. It is affiliated with the University of Montreal for post-graduate instruction. Part of the research programme is biochemical in nature. Daoust and his co-workers, Gaston deLamirande, Réjean Morais, Lionel Poirier, and Vijai Nigam, have studied intensively the metabolism of nucleoproteins and nucleic acids in normal, precancerous, and malignant hepatic tissues, the intracellular distribution of ribonucleases, DNA depolymerase, and phosphomono- and di-esterases, with changes in malignancy. Nigam has devoted his attention mainly to factors in the biosynthesis of glycogen in cells of the Novikoff ascites hepatoma.

Gaston deLamirande was born in Montreal in 1923 and was trained primarily in chemistry at the University of Montreal, where he took his PHD in 1949. He has been associated with the Montreal Cancer Institute since that time.

Biochemistry in Canada in perspective

In the first third of the twentieth century research in biochemistry was carried on in Canada primarily in the universities. Since about 1935 contributions from government laboratories and special institutes have grown in both number and importance. New biochemical knowledge has also come from some departments other than biochemistry, such as chemistry, biology, physiology, nutrition, and experimental medicine. That from dental schools has been negligible.

To attempt to assess the major contributions to biochemical knowledge in Canada over the past seventy years is a difficult task. In my judgment these have been mainly in the fields of endocrinology, neurochemistry, carbohydrate chemistry, and several aspects of phytochemistry. No mention has been made of alkaloids. Their isolation and the determination of their chemical structure have been an outstanding field of Canadian research over the last fifty years carried out mainly by organic chemists such as L. Marion, R.H.F. Manske (editor of *The Alkaloids* in 13 vols.), H.L. Holmes, M. Kulka, W.A. Harrison, O.E. Edwards, D.B. MacLean, K. Wiesner, and Z. Valenta.

ENDOCRINOLOGY

In the early days A. Hunter at Toronto and A.T. Cameron at Manitoba contributed to the relationship of iodine to the thyroid gland; this work was later extended by C.P. Leblond at McGill with the aid of I^{131}. In 1925 J.B. Collip discovered parathormone in parathyroids when at Alberta, and in 1964 D.H. Copp at the University of British Columbia discovered thyrocalcitonin in porcine thyroids. At McGill Collip developed a school of endocrinology between 1928 and

90 The development of biochemistry in Canada

and 1941 which was concerned mainly with the hormones of the anterior pituitary, the placenta, and the ovaries. This was ably carried on there by R.D.H. Heard, J.S.L. Browne, and Eleanor Venning with work on sex hormones and corticosteroids. G.F. Marrian, while in Toronto between 1933 and 1938, with his post-graduate students contributed to our knowledge of the female sex hormones. He isolated oestrone from the urine of pregnant mares and Δ^5-androstenetriol from that of normal males and females. He determined the chemical constitution of equol and studied the constitution and activity of oestriol glucuronide (emmenin).

The discovery of insulin by F.G. Banting and C.H. Best in 1921 at Toronto and its isolation in crude form from pancreas by Collip in 1923 were followed by many biochemical contributions from the departments of Physiology and Medical Research and the Connaught Laboratories at the University of Toronto. One of the most notable was the discovery of zinc in crystalline insulin by D.A. Scott in 1934. A recent accomplishment in this field has been the sulphitolysis of insulin and the recombination of the A and B chains by G.H. Dixon and colleagues at the University of British Columbia in 1962.

Mention should be made here of the studies of D.R. Idler, both at Vancouver and Halifax, of the steroid hormones of fish.

NEUROCHEMISTRY

There have been two major centres of activity in this field: the University of Western Ontario and McGill University-Montreal Neurological Institute. At Western R.J. Rossiter and his students have made many contributions to the metabolism of the phospholipids in brain since 1948 and at McGill K.A.C. Elliott and his colleagues have been concerned with the assay and role of γ-aminobutyric acid since 1944. The metabolism of biogenic amines has been studied by E.A. Hosein at McGill, T.L. Sourkes at the Allan Memorial Institute, and A. D'Iorio at the University of Ottawa since 1956. Some of this work relates to the biochemical basis of epilepsy and the aetiology and treatment of the psychoses.

Since 1937 in the Department of Medical Research at Toronto Erich Baer has studied the synthesis of optically active derivatives of glycerol which led to the synthesis of many phosphatides, cerebrosides, and phosphatidyl peptides. More recently Baer has synthesized numerous phosphatidyl glycols and phosphonolipids.

An extensive programme of investigation of the biosynthesis of the phospholipids of brain and nerve has been developed by R.J. Rossiter during the past twenty years. Extensive use of radioactive or other precursors has been made in this work and synthetic pathways to the phospholipids unravelled. Special atten-

Biochemistry in Canada in perspective 91

tion has been directed recently to the phosphoinositides. At the Collip Research Laboratory in London K.K. Carroll has been investigating various aspects of lipid metabolism such as the distribution of fatty acids in brain and spinal cord and the synthesis and characterization of monoglyceryl esters and ethers of anteiso fatty acids and alcohols.

CARBOHYDRATE CHEMISTRY

Both simple and complex carbohydrates have been the major interest of several Canadian organic chemists, notably: (1) C.B. Purves (1902-65) (52, 53) and his students at McGill University and the Canadian Pulp and Paper Research Institute on cellulose, its oxidation products, and derivatives since 1943; (2) J.K.N. Jones and his students at Queen's University on the biogenesis of plant carbohydrates, especially hemicelluloses and their structures, since 1953; (3) R.U. Lemieux and his associates at the Prairie Regional Laboratory (1949-54), the University of Ottawa (1954-61), and the University of Alberta (1961-) on many aspects of the simple sugars and their stereochemistry; (4) G.A. Adams and C.T. Bishop at the National Research Council in Ottawa on hemicelluloses since 1939, and A.S. Perlin at the Prairie Regional Laboratory especially on fungal polysaccharides since 1949; (5) H.O.L. Fischer and Erich Baer in the Banting and Best Department of Medical Research in Toronto since 1937 on synthetic glyceryl esters and ethers and phospho- and phosphono-lipids.

PHYTOCHEMISTRY

It is appropriate that plant biochemistry should have had a prominent place in Canadian research. As recounted above important contributions have been made from laboratories of the National Research Council at Ottawa, Saskatoon, and Halifax, from institutes of the Department of Agriculture, and from a few universities. Special mention may be made of the fermentation of molasses to produce citric acid and butanediol at the National Research Council in Ottawa from 1940 to 1950; the metabolic activity of the fungus *Ustilago zeae* and the biosynthesis of lignin at the Prairie Regional Laboratory from 1947 to 1960; and the composition and photosynthetic activities of marine algae at the Atlantic Regional Laboratory from 1951 onwards.

Mention should be made of the extensive investigations of cereal grains at the Grain Research Laboratory in Winnipeg since 1933 as directed by W.F. Geddes and J.A. Anderson.

Two university departments have contributed importantly to phytochemical knowledge. G. Krotkov at Queen's from 1950 to 1964 developed chromatographic methods which he applied to aspects of plant metabolism, and A.G. McCalla at Alberta from 1935 to 1945 investigated the proteins of cereal grains.

A relatively new field developing in Canada is the investigation of the composition and products of metabolism of fungi centred mainly at the regional laboratories of the National Research Council and the Institute for Cell Biology of the Department of Agriculture at Ottawa.

Biochemical societies and journals

SOCIETIES

On 10 November 1934, at a joint meeting in Montreal of the Montreal Physiological Society, the Toronto Physiological Society, and the Toronto Biochemical Society, together with representatives from Queen's University and the University of Western Ontario, it was agreed to form a Canadian Physiological Society. J.S.L. Browne was appointed to organize the project and the first annual meeting took place in Toronto on 19 October 1935. A provisional committee was elected to develop the organization, with chairman J. Tait, secretary G.H. Ettinger, treasurer J.K.W. Ferguson, and other members A. Barbeau, R. Blanchet, J.S.L. Browne, A.W. Downs, V.E. Henderson, R.J. Manning, E.W. McHenry, V.H.K. Moorehouse, and E.G. Young. The second annual meeting took place in Kingston on 31 October 1936. The original membership was stated to be 199. Except for 1942-44 meetings have been held annually since 1935.

Many biochemists were members and communicated the results of their research to these annual meetings. In fact, of the eight papers presented at the first meeting in Toronto in 1935, five were biochemical. There is a close parallel between Canadian and American organizations in that the American Society of Biological Chemists was founded in 1906 as an outgrowth of the American Physiological Society which had been founded in 1887. In Great Britain the Biochemical Club (later Society) was started in 1912 and in France the Societé de chimie biologique in 1914.

As the result of the deliberations of an unofficial committee during 1957, consisting of G.C. Butler (chairman), J.M.R. Beveridge, O.F. Denstedt, R. Gau-

TABLE 5
Sequence of officers, location of annual meeting, and membership in the Canadian Biochemical Society

Location	Year	President	Vice-president	Secretary	Treasurer	Membership
Kingston	1957–58	A.M. Wynne	E.G. Young	D.H. Laughland	L.-P. Bouthillier	234
Toronto	1958–59	E.G. Young	O.F. Denstedt	"	"	256
Winnipeg	1959–60	O.F. Denstedt	M. Darrach	"	"	272
Guelph	1960–61	M. Darrach	G.C. Butler	H.D. Branion	M.C. Blanchaer	274
Quebec	1961–62	G.C. Butler	R.J. Rossiter	"	"	271
London	1962–63	R.J. Rossiter	J.H. Quastel	"	R.M. Hochster	274
Halifax	1963–64	J.H. Quastel	L. Berlinguet	"	"	297
Ottawa	1964–65	L. Berlinguet	W.H. Cook	"	P.G. Scholefield	317
Vancouver	1965–66	W.H. Cook	J.A. McCarter	"	"	408
Montreal	1966–67	J.A. McCarter	S.H. Zbarsky	"	J.H. Spencer	520
Kingston	1967–68	S.H. Zbarsky	A. D'Iorio	"	"	600
Edmonton	1968–69	A. D'Iorio	D.R. Whitaker	J.M. Neelin	"	656
Montreal	1969–70	D.R. Whitaker	N.B. Madsen	"	L.-M. Babineau	727
Toronto	1970–71	N.B. Madsen	G.R. Williams	"	"	765
Quebec	1971–72	G.R. Williams	R.H. Hall	N. Begin-Heick	"	807
Saskatoon	1972–73	R.H. Hall	G.E. Connell	"	C. Godin	759
Hamilton	1973–74	G.E. Connell	L.B. Smillie	"	"	
Winnipeg	1974–75	L.B. Smillie	V.P. Sheinin	W. Bridger	"	

dry, D.H. Laughland, E. Pagé, and R.J. Rossiter, a meeting of biochemists was convened on 9 October 1957, in the auditorium of the Medical Building of the University of Ottawa. The chairman was G.C. Butler, and D.H. Laughland acted as secretary. At that meeting it was agreed to form a Canadian Biochemical Society to be affiliated as a founding society within the new Canadian Federation of Biological Societies organized at the same time to include the Canadian Physiological Society, the Canadian Association of Anatomists, the Pharmacological Society of Canada, and the Canadian Biochemical Society. A constitution was drawn up and adopted. The purpose set forth was to foster the science of biochemistry, and the society was to be open to those who, through research or scholarship, had demonstrated an interest in biochemistry. The sequence of annual meetings, officers, and number of members is shown in Table 5. In 1964 an associate membership was opened to post-graduate students in biochemistry or a related field and to graduates of a recognized university employed as biochemists but who were not yet eligible for ordinary membership. In 1974 the total membership was 759 made up of 24 emeritus, 675 ordinary, and 60 student members.

Activities have been confined to communications of original investigations by its members and to the organization of symposia either at the annual meetings or independently. A directory of biochemical training and research in Canada was issued in 1968-69 under the direction of D.R. Whitaker.

Biochemists in Canada have also supported chemical organizations. The Chemical Institute of Canada (CIC) was founded in 1945 as a union of the Canadian Institute of Chemistry founded in 1920, the Canadian Chemical Association founded in 1928, and the Society of Chemical Industry (Canadian Branch) founded in 1902. Within the CIC there has been a biochemical subject division for many years which has held sessions at its annual meetings for the presentation of original communications and has organized symposia, often in collaboration with the Canadian Biochemical Society. Three biochemists have served as president of the CIC: S.A. Beatty in 1950-51, R. Gaudry in 1955-56, and E.G. Young in 1959-60.

The Nutrition Society of Canada was formed in 1957 at a meeting on 8 October at the University of Ottawa with E.W. McHenry as chairman and G.H. Beaton as secretary. This coincided with the founding of the Canadian Federation of Biological Societies but the members of the Nutrition Society were undecided at that time on its proper affiliation and so maintained its independence for several years. Meetings were held in the same place and at about the same time as those of the Federated Societies in 1958, 1959, 1961-63 and of the Agricultural Institute of Canada in 1960 (Table 6). In 1964 the society held a conjoint meeting with the American Society of Nutrition in Toronto on 13-15 September. The society was admitted to membership in the Canadian Federation in 1965.

TABLE 6
Sequence of officers, annual meetings, and membership in the Nutrition Society of Canada

Location	Year	President	Vice-president	Secretary	Treasurer	Membership
Kingston	1958-59	E.W. McHenry	E.W. Crampton	G.H. Beaton	J.A. Campbell	90
Toronto	1959-60	E.W. Crampton	E.H. Bensley	"	"	106
Guelph	1960-61	J. Biely	"	E.V. Evans	"	113
Guelph	1961-62	E.H. Bensley	R.H. Common	"	"	126
Quebec	1962-63	R.H. Common	W.W. Hawkins	"	A.B. Morrison	140
London	1963-64	L.B. Pett	L.B. Pett	"	"	150
Toronto	1964-65	J.M.R. Beveridge	J.M.R. Beveridge	"	"	170
Ottawa	1965-66	G.H. Beaton	G.H. Beaton	K.K. Carroll	"	192
Vancouver	1966-67	J.M. Bell	J.M. Bell	"	E.J. Middleton	192
Montreal	1967-68	J.A. Campbell	J.A. Campbell	R.M. Ballantyne	"	195
Kingston	1968-69	J.M. Demers	J.M. Demers	"	W.E.J. Phillips	197
Edmonton	1969-70	L.E. Lloyd	L.E. Lloyd	"	"	199
Montreal	1970-71	P.J. Lupien	P.J. Lupien	R. Belzile	"	208
Toronto	1971-72	A.B. Morrison	A.B. Morrison	"	J. Beare-Rogers	234
Quebec	1972-73	W.W. Hawkins	W.W. Hawkins	"	"	249
Saskatoon	1973-74	G.J. Brisson	G.J. Brisson	G.R.F. Davis	M. Baigent	265
Hamilton	1974-75	B.E. McDonald	B.E. McDonald	"	"	283

The first constitution sets down the purpose of the society as to extend knowledge of nutrition by the encouragement of research, to provide opportunities for the presentation and discussion of reports of research, to facilitate interchange of nutrition information, and to represent the interests of its members. Membership was open to those individuals who were professionally concerned with scientific aspects of nutrition. In contrast with other Canadian societies concerned with nutrition, emphasis was placed on original research in this field and the number so concerned was comparatively small in 1958, especially in its human aspects in contrast to that of domestic animals.

JOURNALS

The National Research Council of Canada has facilitated the publication of original biochemical research work by sponsoring a series of scientific journals which started as the *Canadian Journal of Research* (1929-35), and was subsequently divided into two sections (1935-43) and then six sections (1944-50), of which E was Medical Sciences. This became the *Canadian Journal of Medical Sciences* (1950-53), then the *Canadian Journal of Biochemistry and Physiology* (1954-63), and finally, in part, the *Canadian Journal of Biochemistry* (1964-). In sequence the editor has been J.B. Collip (1944-56), K.A.C. Elliott (1957-59), J.F. Morgan (1960-67), M.J. Fraser (1967-72), and A. D'Iorio and M. Kates (1973-).

Many biochemical papers have appeared in publications of the Fisheries Research Board: Contributions to Canadian Biology (1901-25), Contributions to Canadian Biology and Fisheries (1925-34), *Journal of the Biological Board of Canada* (1934-37), *Journal of the Fisheries Research Board of Canada* (1938-). A few have been published in the *Revue canadienne de biologie* (1942-) and in the journals of the Agricultural Institute of Canada (1921-).

Distinguished expatriated biochemists

Mention may be made here of several senior biochemists, Canadians by birth or training, who have established international reputations outside of Canada.

Walter Ray Bloor (1877-1965) was born in Ingersoll, Ontario, and educated at Queen's University (MA, 1903) and Harvard (AM, 1908; PHD, 1911). He was professor of biochemistry at the University of Rochester from 1922 to 1947 and devoted most of his life to investigating lipid metabolism.

Earl Judson King (1901-62) (54, 55) was born in Toronto, the son of a Baptist minister. He took his undergraduate course on a scholarship at Brandon College, Manitoba, which was then affiliated with McMaster University. Always highly original in his thinking, Earl earned the money for his education by working as short-order cook in the kitchen-tent of a circus during the summer months. With a BA from McMaster (1923) and PHD in chemistry from Toronto (1926) he joined F.G. Banting in 1926. Because of congestion, King had to do his bench work in the Department of Pathological Chemistry. In return for this hospitality, King helped as a demonstrator in that department and thus acquired experience in this aspect of chemistry, which changed the course of his life. He worked for several years on silicosis, developed micromethods for the determination of silica, and estimated its content in body fluids and tissues.

On leave in 1928-29 he worked with Robert Robison at the Lister Institute in London and at the Kasier-Wilhelm Institute in Munich, where he became familiar with the new microanalytical procedures of F. Pregl. Interest in phosphoric esters acquired from Robison persisted for years as shown by King's many papers on the subject. King improved the colorimetric micromethod for phosphorus, and his procedure is still widely used. He studied phosphatases and with A.R. Armstrong developed an analytical method for their estimation which is still in use.

Late in 1934, King left Toronto for London, where he had been appointed reader in the Department of Chemical Pathology at the British Post-graduate Medical School. He was promoted to the professorship and head of the department in 1945. He continued working on silicosis and other pneumoconioses, and on phosphoric esters and phosphatases, but his main interest became the development of micromethods in clinical chemistry so that 0.1 to 0.2 ml of blood could be used instead of 1, 2, or 5 ml as in earlier methods.

King wrote several books, including *Microanalysis in Medical Biochemistry* in 1946, which has been extensively translated, and one with R.H.S. Thompson on *Biochemical Disorders in Human Disease* in 1957. He deserves much credit for developing clinical biochemistry into a science. He travelled extensively and devoted much time to administrative duties on committees and government service in the United Kingdom, and internationally. His warm uninhibited personality and knowledge of French and German made him a very popular international figure. He was an excellent speaker whose light manner concealed a shrewd brain and strong will. King served for many years on the editorial board of the *Biochemical Journal* and as its chairman in 1946-52 and 1955-59.

Hubert Bradford Vickery (1893-) (56) was born in Yarmouth, Nova Scotia, and educated at Dalhousie University (BSC, 1915; MSC, 1918) and Yale (PHD, 1922). He joined the staff of the Connecticut Agricultural Station at New Haven in 1922 under T.B. Osborne, and became director of the biochemical section in 1928. He retired in 1963. He investigated many aspects of the constituents of plant tissues, especially the metabolism of amino acids and organic acids in tobacco leaves and the composition of plant proteins.

James Murray Luck (1899-) was born in Paris, Ontario, and educated at Toronto (BA, 1922) and Cambridge (PHD, 1925). He has been associated with Stanford University since 1926 and professor of chemistry there since 1939. He was responsible for organizing the *Annual Review of Biochemistry* in 1932 and served as editor of thirty-four volumes until he retired in 1965.

Dilworth Wayne Woolley (1914-66) was born in Raymond, Alberta, and died in Cuzco, Peru. He obtained his BSC (1935) from the University of Alberta and the MSC (1936) and PHD (1938) from the University of Wisconsin. He joined the staff of the Rockefeller Institute in New York in 1939 and remained there for the rest of his life. Despite the handicap of blindness he made several outstanding contributions to biochemistry. He isolated and identified nicotinic acid as the nutritional deficiency in pellegra, he established the structure of pantothenic acid by synthesis, detected serotonin as an agent in mental disease, and established inositol as a member of the B vitamins. His major interest lay in the mode of action of vitamins, and he must be regarded as a very distinguished biochemist.

Thomas Hughes Jukes was born in Hastings, England, in 1906. He attended the University of Toronto and was awarded a doctorate in biochemistry under

H.D. Kay in 1933 when he determined the distribution of amino acids in livetin of hen's egg yolk. As a National Research Council Fellow he worked at the University of California (1933-34). He then worked in the Institute of Poultry Husbandry (1934-39) and taught on the staff of the University of California (1939-42). He became director of the section on nutrition of the Lederle Laboratories (1942-59), but returned to the University of California (Berkeley) as professor in residence of medical physics and, from 1963 to the present, as research biochemist in the Space Science Laboratory. He has made noteworthy contributions in the field of nutrition, especially concerning the B vitamins, pantothenic and folic acids, and the effect of antibiotics.

Reginald MacGregor Archibald was born in 1910 in Syracuse, New York. He received his university education in Canada (BA at UBC, 1930; MA, 1932, and PHD, 1934, at Toronto). His doctorate at Toronto was in pathological chemistry under V.J. Harding. He qualified for the MD degree in 1939, and, after internships at the Hospital for Sick Children and the Toronto General, he served as resident at the Hospital of the Rockefeller Institute in New York from 1941 to 1946. Archibald was then appointed professor and head of the Department of Biochemistry at Johns Hopkins in 1946, but returned to the Rockefeller in 1948. He was promoted to professor of medicine and senior physician at the hospital in 1959, which offices he still holds.

His research work delves into many fields: the non-fermentable carbohydrates and steroids in urine, the physiological role of glutamine, the influence of hormones on enzymic activities, the growth of children with skeletal anomalies.

Other distinguished, expatriated but younger native Canadians include Samuel Kirkwood, professor of biochemistry at the University of Minnesota; F.R.N. Gurd, professor of biochemistry at Indiana University; G.H. Lathe, professor of chemical pathology at the University of Leeds; J.B. Neilands, professor of biochemistry at the University of California (Berkeley); M.D. Kamen, born in Toronto in 1913 but educated at the University of Chicago, professor of chemistry at the University of California (San Diego); Julius Marmur, born in Poland but educated at McGill, professor of biochemistry at the Albert Einstein College of Medicine in New York; Harry Rudney, born and educated at Toronto, professor of biochemistry and head of the department in the Medical College of the University of Cincinnati; and probably others of whom I am not aware.

Financial support of biochemical research

It has recently been estimated that there are about 1500 biochemists now engaged in teaching, research, or other activities in Canada (57). In 1966 monetary expenditures on biochemical research were subdivided as follows:

	$	%
Federal government (intramural)	2,510,000	24.1
Provincial governments	0	0
Industry	1,786,000	17.2
Universities and research institutes	6,129,000	58.7
TOTAL	10,425,000	100.0

Such figures, however, must be taken as very approximate because of the difficulty of classifying biochemical research. Canadian sources of financial support as grants-in-aid are listed below (many are restricted to specific fields of research):

National Research Council since 1917
Banting Research Foundation since 1925
National Cancer Institute since 1947
Defence Research Board since 1947
Department of National Health since 1948

TABLE 7
Grants-in-aid of research from Medical Research Division or Council
(including those for equipment only)

Year	Total of all grants-in-aid (dollars)	Grants to Depts. of Biochem. and Path. Chem. (dollars)	Percentage of total	Number of projects
1955–56	652,056			
1956–57	701,126			
1957–58	745,592			
1958–59	1,321,379	256,135	19.4	45
1959–60	1,723,846	250,450	14.5	37
1960–61	1,899,809	284,960	15.0	42
1961–62	2,673,456	448,038	16.8	56
1962–63	3,428,556	528,368	15.4	64
1963–64	3,963,911	583,018	14.7	67
1964–65	5,177,888	823,342	15.9	70
1965–66	7,133,810	888,893	12.5	76
1966–67	12,087,607	1,461,052	12.1	109
1967–68	15,388,049	2,213,410	14.3	130
1968–69	19,232,023	2,719,752	14.2	154
1969–70	20,770,652	2,785,654	13.4	161
1970–71	22,478,848	2,901,091	12.9	161

Canadian Arthritis and Rheumatism Society since 1948
Multiple Sclerosis Society of Canada since 1948
Muscular Dystrophy Association of Canada since 1948
J.P. Bickell Foundation since 1951
Canadian Heart Foundation since 1956
Canadian Association for the Mentally Retarded since 1958
Canadian Cystic Fibrosis Foundation since 1960
Medical Research Council since 1960 (as Medical Research Committee or Division of NRC, 1938–60)

Of these, the major sources of financial support have been the National Research Council, now as the Medical Research Council, and the National Cancer Institute, with very variable and fewer grants from the others (58). Prior to World War II grants from the National Research Council for biochemical research, other than intramural, were rather insignificant and this remained true even up to 1958, as can be seen in Table 7. Since the establishment of the Medical Research Council in 1960 the grants have increased significantly. The figures of grants quoted are undoubtedly too low because they represent only grants to individuals in departments of biochemistry and not biochemical projects covered by block or other grants to institutes. The allotment of the total grants-in-aid assigned to biochemistry has been approximately 15 per cent, and the total has reached about $3,000,000 annually. To this should be added those grants made to individual biochemists by the National Research Council as distinct from the Medical Research Division or Council classified either under biology or chemistry. Such additions are now not large proportionally but were more significant ten years ago.

Technical progress

It is probably very difficult for the young biochemist to appreciate the limitations of the tools available in the early part of the twentieth century in contrast with those available today. The writer graduated from an honours course in chemistry and biology in 1916. He can thus contrast the technology of that period with current practice, an interval of over fifty years.

The purity of commercial chemicals was uncertain in those days and recrystallization was frequently necessary. The German companies of Kahlbaum and Merck were preferred and did supply a special grade *für Analyse*. The J.T. Baker Chemical Co. of Phillipsburg was just introducing the policy of supplying chemicals with an analysis (1904). The term CP (chemically pure) was used generally but no standards had been formulated so that it had little meaning. Organic chemicals were manufactured by many small companies and were often difficult to obtain. Eastman Kodak Co. of Rochester had not yet begun to function as a central supply house. One often had to make them oneself with the help of Beilstein or Houben.

There were four general scientific supply houses in North America: Eimer and Amend of New York (1851), Central Scientific of Chicago (1889), Arthur H. Thomas of Philadelphia (1900), and Fisher Scientific of Pittsburg (1902). The Canadian branch of the Fisher Scientific Co. was started in Montreal in 1922, Canadian Laboratory Supplies in 1923, and Central Scientific Co. of Canada in 1926.

The borosilicate glass made by Corning (patented in 1915) was only beginning to appear as Pyrex and was much more expensive than the ordinary. Direct heating with Bunsen or Meker burners was not possible, and many accidents occurred with vessels of the fragile Bohemian or even Jena glass.

Porcelain Buchner funnels and graded Whatman or Schleicher and Schüll filter papers were in common use with vacuum supplied by individual glass pumps of very variable efficiency (as is still true). Vacuum desiccators did occasionally collapse.

The most sensitive chemical balance was the Beker or Sartorius with two pans and weights that had to be calibrated. It was sensitive to about 0.2 mg. The Kuhlmann microbalance with sensitivity of 0.001 mg as modified by Pregl was rare and only became commercially available in 1924 as a modification of the earlier assay balance.

Small unguarded centrifuges for bench use were common and used especially in the analysis of milk as in the Babcock method for fat and for the centrifugation of blood and urine. Large models, like the International No. 1 and 2, were less common and the glass vessels frequently broke at 2000 rpm.

Drying was done in Freas electrically heated ovens without vacuum. The muffle furnace was available for ashing but this was more often done with the open flame.

Dialysis required the preparation of membranes from collodian (cellulose nitrate in ethanol-ether) which one made oneself from cotton wool. The time of drying regulated the permeability of the membrane.

The importance of the measurement of pH was just being realized, and the Clark and Lubs buffers and indicators were popular. *The Determination of Hydrogen Ions* by W.M. Clark was first published in 1920. The hydrogen electrode had just been invented and you made it yourself if you required one. A supply of gaseous hydrogen from a Kipp generator was required. Buffer salts all had to be recrystallized. The commercial glass electrode was not available until 1935.

Spectrometry in the infrared or ultraviolet had not touched biochemistry but the hand spectroscope was in use, especially for the study of haemoglobin and its derivatives. The dipping refractometer was used for fats and oils and the polarimeter (with the mercury arc lamp or more commonly the sodium flame) was a common instrument, especially for sugars.

The Duboscq colorimeter was becoming popular but it required simultaneous visual comparison with the standard. The Lovibond tintometer was devised in 1910 but one was rarely seen in Canada before 1925. Nessler tubes of 75 or 100 ml were commonly used in colorimetry, especially for water analysis. The photoelectric colorimeter was invented about 1940 (e.g. the Canadian one of K.A. Evelyn [q.v.]).

The 'bomb' calorimeter had come into use for the determination of the energy value of foods. A very few respiration calorimeters had been built for use with human subjects or animals (e.g. the Atwater-Rosa of 1897).

Gas analysis was done with the clumsy Hempl burette and pipettes or the more compact Orsat apparatus. These are to be contrasted with the more modern manometric methods of Haldane-Henderson, Van Slyke, and Warburg.

The whole field of chromatography only began to evolve with the work of Martin and Synge in 1941 on liquid-liquid separations, and of James and Martin in 1952 on gas-liquid separations. The application of mass spectrometry to biochemistry is very recent. Electrophoretic methods in biochemistry, however, date from the work of Tiselius in 1937 at Uppsala, when he constructed a three-sectional U tube to operate at $1°c$ in the moving boundary technique. The first such instruments, as the Longsworth model, to be installed in Canada were at Dalhousie and Alberta around 1948. One at the National Research Council in Ottawa was assembled in 1950.

The ultracentrifuge has changed from that of Svedberg (1924) – a monster which was driven by oil-turbine, required two floors, and of which few existed – to the compact electrical instruments of Spinco. The first analytical ultracentrifuge in Canada was a Spinco Model E installed in the Division of Applied Biology at the National Research Council in Ottawa in 1949 at a cost of $20,000. (There are many such instruments in Canada today, although the cost is now over $25,000.)

Analytical methods have played a dominant role in chemical research. This is well illustrated by the Kjeldahl method for organic nitrogen and the Soxhlet apparatus for lipids, which have changed little in fifty years while methods of estimating amino and fatty acids have been revolutionized. The writer recalls spending two weeks in 1938 to analyse ichthulokeratin for arginine, lysine, and histidine – gravimetrically – and two days in 1956 to analyse all eighteen amino acids present by the Moore-Stein chromatographic method.

Otto Folin was largely responsible for the development of analytical methods applied to urine dating from about 1905 and for the classic Folin-Wu system in blood of 1919. Many of his methods are still in active use but automatized in hospital service.

Many sugars were recognized – mainly by their osazones – and estimated by their reducing power, volumetrically or gravimetrically.

The first enzyme to be crystallized was urease. This was achieved by Sumner in 1926, and led to the subsequent controversy over the protein nature of enzymes. Now new enzymes are crystallized every month.

A vitamin was first discovered by Hopkins in 1912. It was then called an accessory food factor. The twentieth century has thus seen a great development in the subject of nutrition as exemplified by the four editions of Graham Lusk's *Science of Nutrition* from 1906 to 1928 and quantitation of nutrients, especially vitamins and minerals, in both raw and cooked foods.

The first edition of Hawk's *Practical Physiological Chemistry* was published in 1907 and contained about 200 pages. The 14th and last edition appeared in 1965 with 1500 pages. One may also contrast the one-volume textbook of 1900 of about 500 pages (Mathews, Hammarstein or Abderhalden) with the current *Comprehensive Biochemistry*, edited by Florkin and Stotz, which consists of 28 volumes, each of about 500 pages.

In 1910 there were only four biochemical journals - *Journal of Biological Chemistry* (1905), *Biochemical Journal* (1906), *Biochemische Zeitschrift* (1906), and *Zeitschrift für physiologische Chemie* (1877) - and one could read almost every article with some degree of comprehension as L.B. Mendel did as a graduate student with the *J.B.C.* Those days are gone forever.

References

1 Chittenden, R.H., *The Development of Physiological Chemistry in the United States* (Chem. Catalog Co., New York, 1930)
2 Macallum, A.B., 'The Development of Physiology and Biochemistry in Canada,' in *Fifty Years Retrospect*, Roy. Soc. Canada Anniversary Vol., 1882-1932, pp. 163-5 (1932)
3 Flexner, A., *Medical Education in the United States and Canada* (Carnegie Found. Bull. No. 4, New York, 1910)
4 Leathes, J.B., Obituary of A.B. Macallum, *Roy. Soc. (London) Obituary Notices* 1: 287 (1932-35)
5 Obituary of A.B. Macallum, *Trans. Roy. Soc. Canada*, 3rd Ser., 28: xix (1934)
6 Collip, J.B., Obituary of A.T. Cameron, *Proc. Roy. Soc. Canada*, 3rd Ser., 42: 83 (1948)
7 Macallum, A.B., Obituary of R.F. Ruttan, *Trans. Roy. Soc. Canada*, 3rd Ser., 24: vii (1930)
8 Obituary of V.J. Harding, *Trans. Roy Soc. Canada*, 3rd Ser., 29: iv (1935)
9 Haworth, W.N., Obituary of V.J. Harding, *J. Chem. Soc.* 1341 (1935)
10 King, E.J., 'Victor John Harding,' *Biochem. J.* 29: 1 (1935)
11 Elliott, K.A.C., Obituary of D.L. Thomson, *Proc. Roy. Soc. Canada*, 4th Ser., 3: 177 (1965)
12 Clark, R.H., Obituary of Harold Hibbert, *Proc. Roy. Soc. Canada*, 3rd Ser., 40: 95 (1946)
13 Rossiter, R.J., Obituary of J.B. Collip, *Proc. Roy. Soc. Canada*, 4th Ser., 4: 73 (1966)

14 Browne, J.S.L., and Denstedt, O.F., 'J.B. Collip,' *Endocrinology* 79: 225 (1966)
15 Thomson, D.L., 'Dr. J.B. Collip,' *Can. J. Biochem. Physiol.* 35, Nov. (1957)
16 Barr, M.L., and Rossiter, R.J., 'James Bertram Collip,' *Biographical Memoirs of Fellows of the Roy. Soc.* 19: 235 (1973)
17 Obituary of T.B. Robertson, *Biochem. J.* 24: 577 (1930)
18 Marston, H.R., 'Obituary and Bibliography of T.B. Robertson,' *Australian J. Exptl. Biol. Med. Sci.* 9: 1 (1932)
19 Dauphinee, J.A., Obituary of Andrew Hunter, *Proc. Roy. Soc. Canada*, 4th Ser., 8: 83 (1970)
20 Wynne, A.M., Obituary of H. Wasteneys, *Proc. Roy. Soc. Canada*, 4th Ser., 5: 125 (1967)
21 Scott, D.A., Obituary of E.W. McHenry, *Proc. Roy. Soc. Canada*, 3rd Ser., 56: 219 (1962)
22 Beveridge, J.M.R., 'The History of Biochemistry at Queen's,' *Queen's Rev.* 29: 192 (1954)
23 Ettinger, G.H., Obituary of R.G. Sinclair, *Proc. Roy. Soc. Canada*, 3rd Ser., 44: 105 (1950)
24 Evans, C.L., Obituary of J.B. Leathes, *Nature* 178: 833 (1956)
25 Graham, D., Obituary of J.B. Leathes, *Proc. Roy. Soc. Canada*, 3rd Ser., 52: 89 (1958)
26 Peters, R.A., 'J.B. Leathes,' *Biographical Memoirs of Fellows of the Roy. Soc.* 4: 185 (1958)
27 Ferguson, J.K.W., Obituary of D.A. Scott, *Proc. Roy. Soc. Canada*, 4th Ser., 10: 93 (1972)
28 Best, C.H., and Fisher, A.M., 'David Alymer Scott,' *Biographical Memoirs of Fellows of the Roy. Soc.* 18: 511 (1972)
29 Thistle, M., *The Inner Ring* (University of Toronto Press, Toronto, 1966)
30 National Research Council, Annual Reports and Reviews (Queen's Printer, Ottawa, 1932-70)
31 *History of the Wartime Activities of the Division of Applied Biology.* (NRC, Ottawa, n.d.)
32 Eggleston, W., *Scientists at War* (University of Toronto Press, Toronto, 1950)
33 Carter, N.M., Obituary of H.N. Brocklesby, *Proc. Roy. Soc. Canada*, 4th Ser., 2: 83 (1964)
34 Davidson, A.L., *The Genesis and Growth of Food and Drug Administration in Canada* (Department of National Health and Welfare, Ottawa, 1949)
35 Warrington, C.J.S., and Nicholls, R.V.V., *A History of Chemistry in Canada* (Pittman and Sons Ltd., Toronto, 1949)
36 *National Health Review*, Jan. 1940

37 *Can. Bull. Nutr.* 2: 1 (1950)
38 *Can. Bull. Nutr.* 6: 1 (1964)
39 Hawkins, W.W., *Can. J. Pub. Health* 61: 196 (1970)
40 *Nutrition: A National Priority* (Information Canada, Ottawa, 1973)
41 Brown, A., 'Frederick F. Tisdall,' *Can. Med. Assoc. J.* 64: 263 (1951)
42 Aitken, T.R., *The Board of Grain Commissioners for Canada: Grain Research Laboratory*, 1913-1938 (1962)
43 Anderson, J.A., *Cereal Chem.* 34, No. 6: Suppl. 1-37 (1957)
44 Wilson, A.H., *Research Councils in the Provinces - A Canadian Resource* (Science Council of Canada, Special Study No. 19; Queen's Printer, Ottawa, 1971)
45 Cook, W.H., Obituary of N.H. Grace, *Proc. Roy. Soc. Canada*, 3rd Ser., 56: 187 (1962)
46 Best, C.H., *Can. Med. Assoc. J.* 87: 1046 (1962)
47 Hunter, A., Obituary of F.G. Banting, *Proc. Roy. Soc. Canada*, 3rd Ser., 35: 87 (1941)
48 Best, C.H., Obituary of F.G. Banting, *Roy. Soc. (London) Obituary Notices* 4: 21 (1942)
49 Best, C.H., *J. Endocrinol.* 19: i (1959)
50 Sowden, J.C., 'H.O.L. Fischer,' *Adv. Carbo. Chem.* 17: 1 (1962)
51 Fischer, H.O.L., *Ann. Rev. Biochem.* 29: 1 (1960)
52 Perlin, A.S., 'C.B. Purves (1902-1965),' *Adv. Carbo. Chem.* 23: 1 (1968)
53 Winkler, C.A., 'C.B. Purves (1902-1965),' *Proc. Roy. Soc. Canada*, 4th Ser., 4: 111 (1966)
54 Neuberger, A., and Klyne, W., 'Earl J. King,' *Biochem. J.* 89: 401 (1963)
55 Harrison, C.V., 'Earl J. King,' *J. Path. Bact.* 88: 601 (1964)
56 Vickery, H.B., *Ann. Rev. Plant Physiol.* 23: 1 (1972)
57 Science Council of Canada, Special Study No. 9, *Chemistry and Chemical Engineering - A Survey of Research and Development in Canada* (Queen's Printer, Ottawa, 1969)
58 Reference Lists of Health Science Research in Canada, 1960-70 (Medical Research Council, Ottawa)

APPENDIX
List of biochemists cited, with academic qualifications*

Ackman, R.G., BA (Tor.) 50, MSc (Dal.) 52, PhD (London) 56
Adams, G.A., BA (Queen's) 28, MSc (UWO) 30, PhD (Chicago) 38, FRSC 53
Allardyce, W.J., BA (UBC) 19, MA 21, PhD (McGill) 31
Anderson, J.A., BSc (Alta.) 26, MSc 28, PhD (Leeds) 30, FRSC 44
Archibald, R.M., BA (UBC) 30, MA 32, PhD (Tor.) 34, MD 39

Babineau, L.-M., BA (Laval) 40, BSc 45, DSc 55
Baer, Erich, PhD (Berlin) 27, FRSC 56
Baril, G.H.,[†] BA (Montreal) 04, MD (Laval) 08
Bayley, S.T., BSc (London) 46, PhD 50
Beall, D., BA (UBC) 32, PhD (Tor.) 36
Beaton, G.H., BA (Tor.) 52, MA 53, PhD 55
Beatty, S.A., BA (Queen's) 25, MA 26, PhD (McGill) 29
Begg, R.W., BSc (Dal.) 36, MSc 38, MD 42, DPhil (Oxon.) 50, FRCP(C)
Belleau, B.R., BSc (Montreal) 47, MSc 48, PhD (McGill) 50, FRSC 68
Benoiton, N.L., BSc (Loyola) 53, MSc (Montreal) 55, PhD 56
Benson, C.C.,[†] BA (Tor.) 99, PhD 03
Berlinguet, L., BSc (Montreal) 47, PhD (Laval) 50, FRSC 69
Best, C.H., BA (Tor.) 21, MA 22, MD 25, DSc (London) 28, FRSC 31, FRS 38, FRCP(C), CBE, CH
Beveridge, J.M.R., BSc (Acadia) 37, PhD (Tor.) 40, MD (UWO) 50, FRSC 60

*Degrees, honoris causa, omitted.
[†] Deceased.

Biely, J., BSA (UBC) 26, MSA 30, MS (Kansas State) 29, FRSC 66
Birchard, F.J.,[†] BA (Tor.) 01, PhD (Leipzig) 09
Bishop, C.T., BSc (Acadia) 45, BA 46, PhD (McGill) 49, FRSC 72
Blanchaer, M.C., BA (Queen's) 45, MD, CM 46
Bligh, E.G., BSc (Acadia) 49, MSc 51, PhD (McGill) 56
Bliss, S.,[†] BSc (Ohio State) 17, PhD (Harvard) 25
Bloor, W.R.,[†] BA (Queen's) 02, AM (Harvard) 08, PhD 11
Bois, E., BA (Laval) 20, LSc 24, DSc (Freibourg) 27
Bouthillier, L.-P., BSc (Montreal) 35, MSc 36, PhD (Ill.) 45
Branion, H.D., BA (Tor.) 28, MA 29, PhD 33
Brocklesby, H.N.,[†] BSc (Man.) 26, MSc 27, PhD (McGill) 35, FRSC 39
Browne, J.S.L., BA (McGill) 26, BSc 27, MD 29, PhD 32, FRSC 39, FRCP(C)
Butler, G.C., BA (Tor.) 35, PhD 38, FRSC 57

Cameron, A.T.,[†] MA (Edin.) 04, BSc 06, DSc 25, FRSC 20, CMG
Campbell, J.A., BSA (Tor.) 36, MSc (McGill) 38, PhD 47
Campbell, W.R., BA (Tor.) 11, MA 12, MB 15, MD, FRSC 33, FRCP(C)
Cantero, A., BA (Ottawa) 27, MD (McGill) 27, FRSC 55
Carroll, K.K., BSc (UNB) 43, MSc 46, MA (Tor.) 46, PhD (UWO) 49
Carter, N.M., BASc (UBC) 25, MASc 26, PhD (McGill) 29
Chan, A.P., BSc (McGill) 44, MSc 46, PhD (Ohio State) 49
Chapman, R.A., BSA (Tor.) 40, MSc (McGill) 41, PhD 44
Charles, A.F., BA (Tor.) 27, MA 28, PhD 32
Cinader, B., BSc (London) 45, PhD 49, DSc 58, FRSC 71
Cipriani, A.J.,[†] BSc (McGill) 32, MD 40, FRSC 54
Cohen, S.L., BA (Brandon) 32, PhD (Tor.) 36
Collier, H.B., BA (Tor.) 27, MA 29, PhD 30, FRSC 54
Collip, J.B.,[†] BA (Tor.) 12, MA 13, PhD 16, DSc (Alta.) 24, MD 26, FRSC 25, FRS 33, CBE
Colter, J.S., BSc (Alta.) 45, PhD (McGill) 51, FRSC 72
Colvin, J.R., BSA (Sask.) 46, MSc (Alta.) 48, PhD (Minn.) 51
Common, R.H., BSc (Belfast) 28, BAgr 29, MAgr 31, DSc 57, PhD (London) 35, DSc 44, FRSC 65
Connell, G.E., BA (Tor.) 51, PhD 55
Cook, A.S.,[†] BA (Dal.) 26, MA 27, PhD (Tor.) 31
Cook, W.H., BSc (Alta.) 26, MSc 28, PhD (Stanford) 31, FRSC 43, OBE
Craig, B.M., BSc (Sask.) 44, MSc 46, PhD (Minn.) 50
Crampton, E.W., BS (Conn. Agr. Coll.) 19, MS (Iowa State) 22, PhD (Cornell) 37, FRSC 45
Crocker, B.F., BA (Tor.), MA 33, PhD 40

List of biochemists cited 115

Darrach, M., BA (UBC) 35, MA 36, PhD (Tor.) 40
Dauphinee, J.A., BA (UBC) 22, MA 23, PhD (Tor.) 29, MD 30, FRSC 53, FRCP(C), OBE
deLamirande, G., BA (Montreal) 43, BSc 46, MSc 47, PhD 49
Delory, G.E., BSc (London) 39, MSc 42, PhD 45
Denstedt, O.F.,[†] BSc (Man.) 29, PhD (McGill) 37, FRSC 64
Depocas, F., BSc (Montreal) 46, PhD 51
Despointes, R.H.,[†] BSc (Caen) 44, MD (Paris) 50, PhD (McGill) 62
D'Iorio, A., BSc (Montreal) 46, PhD 49, FRSC 69
Dixon, G.H., BA (Cantab.) 51, PhD (Tor.) 56, FRSC 70
Dvornik, D.M., Chem. Eng. Zagreb, 48, PhD 54
Dyer, W.J., BSc (St F.X.) 34, MSc (McGill) 37, PhD 40

Eagles, B.A., BA (UBC) 22, MA (Tor.) 24, PhD 26, FRSC 52
Elliott, K.A.C., BSc (Rhodes) 23, MSc 24, PhD (Cantab.) 30, ScD 50, FRSC 63
Emslie, A.R.G.,[†] BSA (OAC) 28, MSA (Tor.) 31, DSc (Aberdeen) 34
Evelyn, K.A., BSc (McGill) 32, MD 38

Feltham, L.A.W., BSc (Dal.) 47, MA 49, PhD (Tor.) 60
Finn, D.B., BSc (Man.) 24, MSc 28, PhD (Cantab.) 33
Fischer, H.O.L.,[†] PhD (Jena) 12
Fisher, A.M., BA (Tor.) 31, MA 32, PhD 34
Fisher, J. Manery, BA (Tor.) 32, MA 33, PhD 35
Fraser, M.J., BSc (Dal.) 52, MSc 54, PhD (Cantab.) 57

Gaebler, O.H., AB (Central Wesleyan Coll.) 17, AM (Missouri) 20, PhD (Tor.) 22, MD (Cornell) 31
Gaudry, R., BA (Laval) 33, BSc 37, DSc 40, FRSC 54, CC
Geddes, W.F.,[†] BSA (OAC) 18, MA (Tor.) 25, MSc (Minn.) 28, PhD 29
Gianetto, R., BSc (Montreal) 49, MSc 51, PhD 53
Gingras, J.R., BA (Laval) 29, MD 34
Godin, C., BA (Laval) 46, BSA 50, DSc 53
Gorham, P.R., BA (UNB) 38, MSc (Me.) 40, PhD (Cal. Tech.) 43
Gornall, A.G., BA (Mt A.) 36, PhD (Tor.) 41, FRSC 66
Grace, N.H.,[†] BA (Sask.) 25, MA 27, PhD (McGill) 31, FRSC 48, MBE
Graham, A.F., BASc (Tor.) 38, MASc 39, PhD (Edin.) 42, DSc 52
Graham, W.R., Jr., BSA (OAC) 29, MSA (Tor.) 31, PhD 33
Grant, G.A.,[†] BSc (Dal.) 27, MSc 29, PhD (Tor.) 32 and (London) 37
Gurd, F.R.N., BSc (McGill) 45, MSc 46, PhD (Harvard) 49

Hagen, P.B., MB, BS (Sydney) 45
Hall, R.H., AB (Syracuse) 31, MS (NYU) 35, PhD (Columbia) 41
Hanes, C.S., BA (Tor.) 25, PhD (Cantab.) 31, ScD 51, FRS 42, FRSC 56
Harding, V.J.,[†] BSc (Manchester) 06, MSc 08, DSc 12, FRSC 23
Hart, J.S.,[†] BA (Tor.) 39, MA 42, PhD 49, FRSC 59
Haskins, R.H., BA (UWO) 38, MA 40, PhD (Harvard) 48
Hawkins, W.W., BA (UNB) 39, MSc (Dal.) 43, PhD (Tor.) 51
Hayes, F.R., BSc (Dal.) 26, MSc 27, PhD (Liverpool) 29, DSc 48, FRSC 47
Heard, R.D.H., BA (Tor.) 29, MA 30, PhD (Manchester) 32, FRSC 49
Helleiner, C.W., BA (Tor.) 52, PhD 55
Henderson, J.F., BSc (Ariz.) 54, MSc 56, PhD (Wisc.) 59
Hibbert, H.,[†] BA (Manchester) 97, MA 00, DSc 11, PhD (Leipzig) 06, FRSC 31
Hlynka, I., BSc (Alta.) 35, MSc 37, PhD (Cal. Tech.) 40
Hochster, R.M.,[†] BA (Sir G.W.) 46, PhD (McGill) 50
Hosein, E.A., BSc (McGill) 47, MSc 50, PhD 52
Hunter, A.,[†] MA (Edin.) 95, BSc 99, MB, ChB 01, FRSC 16, FRSE 32, CBE
Hunter, G., MA (Glasgow) 18, BSc 21, DSc 29, FRSC 33
Hurst, R.O., BA (Tor.) 41, PhD 52

Idler, D.R., BA (UBC) 49, MSc 50, PhD (Wisc.) 53, FRSC 72
Irvine, G.N., BSc (Man.) 43, PhD (McGill) 49

Jackson, S.H., BSc (Tor.) 32, MA 34, PhD 36
Jean, M., BA (Laval) 37, BSc 41, DSc 52
Jellinck, P.H., BA (Cantab.) 48, BSc (London) 50, MSc 52, PhD 54
Jones, J.K.N., BSc (Birm.) 33, PhD 36, DSc 49, FRS 57, FRSC 60
Jukes, T.H., BSA (OAC) 30, PhD (Tor.) 33

Kamen, M.D., BS (Chicago) 33, PhD 36
Kaneda, T., BEng (Tokyo) 50, DSc 62
Kates, M., BA (Tor.) 45, MA 46, PhD 48, FRSC 72
Kay, C.M., BSc (McGill) 52, PhD (Harvard) 56
Kay, H.D.,[†] BSc (Manchester) 14, MSc 23, DSc 26, PhD (Cantab.) 25, FRS 45, CBE
Khorana, H.G., BSc (Punjab) 43, MSc 45, PhD (Liverpool) 48
King, E.J.,[†] BSc (McMaster) 23, MA 24, PhD (Tor.) 26
Kirkwood, S., BSc (Alta.) 42, MSc (Wisc.) 44, PhD 47

Labarre, J., BPh, LèsS (Laval), DSc (Paris), FRSC 45
Labrie, F., BA (Laval) 57, MD 62, PhD 67

Lachance, J.P., BA (Lévis) 45, BSc (Laval) 49, PhD 53
Laidler, K.J., BA (Oxon.) 37, DSc 56, PhD (Princeton) 40, FRSC 60
Lamy, F., Lic ès Sci (Paris) 48, MA (Amherst) 50, PhD (MIT) 55
Larmour, R.K.,[†] BSc (Sask.) 23, MSc 25, PhD (Minn.) 27, FRSC 46
Lathe, G.H., BSc (McGill) 34, MSc 36, PhD 47, MD 38
Laughland, D.H., BSA (OAC) 38, MA (Tor.) 42, PhD 49
Layne, D.S., BSc (McGill) 53, MSc 55, PhD 57
Leathes, J.B.,[†] MB (Guy's) 93, FRCS 94, FRS 11, FRCP 21
Lemieux, R.U., BSc (Alta.) 43, PhD (McGill) 46, FRSC 57, FRS 67
Lloyd, L.E., BSA (McGill) 48, MSc 50, PhD 52
Logan, J.F.,[†] BA (Acadia) 13, MA 14, AM (Yale) 16, PhD (McGill) 23
Lothrop, A.P.,[†] AB (Oberlin) 06, AM 07, PhD (Columbia) 09
Lozinski, E., MD (McGill) 20, MSc 23
Lucas, C.C., BA (UBC) 25, MASc 26, PhD (Tor.) 36, FRSC 59
Luck, J.M., BA (Tor.) 22, PhD (Cantab.) 25
Lupien, P.J., BSc (Ottawa) 54, MSc 56, PhD (Cornell) 66

Macallum, A. Bruce, BA (Tor.) 07, MB 09, MD 10, PhD 19, FRSC 32
Macallum, A. Byron,[†] BA (Tor.) 80, PhD (Johns Hopk.) 88, MB (Tor.) 90, FRSC 00, FRS 06
MacPherson, C.F.C., BSc (Mt A.) 35, MSc (Dal.) 37, PhD (Columbia) 45
Macpherson, L.B., BSc (Acadia) 34, PhD (Tor.) 49, MBE
MacRae, H.F., BSc (McGill) 54, MSc 56, PhD 60
McArthur, C.S., BA (UWO) 35, MSc 38, PhD (Tor.) 43
McCalla, A.G., BSc (Alta.) 29, MSc 31, PhD (Cal.) 33, FRSC 53
McCarter, J.A., BA (UBC) 39, PhD (Tor.) 45, FRSC 64
McConnell, W.B., BSc (Alta.) 46, MSc 47, PhD (McGill) 49
McFarlane, W.D.,[†] BSc (Tor.) 25, MA 29, PhD 32, FRSC 42
McGill, A.,[†] BA (Tor.) 80, BSc 82, FRSC 00
McHenry, E.W.,[†] BA (Tor.) 21, MA 23, PhD 29, FRSC 42
McLean, W.F., BASc (Tor.) 37
Madsen, N.B., BSc (Alta.) 50, MSc 52, PhD (Wash.) 55
Manning, R.J., BSc (Tor.) 15, DSc (Bristol)
Marko, A.M., BA (Sask.) 46, MD (Tor.) 49, PhD 52
Marmur, J., BSc (McGill) 46, MSc 47, PhD (Iowa State) 51
Marrian, G.F., BSc (London) 25, DSc 30, FRSC 37, FRS 44, FRSE
Martin, H., BSc (London) 21, MSc 25, DSc 34
Martin, W.G., BSc (Carleton) 52, MSc (McGill) 55, PhD 58
Mawson, C.A., BSc (Manchester) 29, MSc 30, PhD 32
Middleton, E.J., BSc (UWO) 52, MSc 55, PhD (Rutgers) 59

Migicovsky, B.B., BSA (Man.) 35, PhD (Minn.) 40
Moloney, P.J., BA (Tor.) 12, MA 15, PhD 24, FRSC 36, OBE
Morgan, J.F., BA and BSA (UBC) 41, MSA 42, PhD (Tor.) 46, FRSC 61
Morrell, C.A., BA (Tor.) 24, MA 25, PhD (Harvard) 30, FRSC 47
Morrison, A.B., BSc (Alta.) 51, MSc 52, PhD (Cornell) 56
Murray, T.K., BSc (McGill) 48, MSc 50, PhD 57
Murthy, M.R.V., BSc (Mysore) 49, PhD (Bangalore) 55

Neelin, J.M., BA (Tor.) 53, PhD 58
Neilands, J.B., BSc (Tor.) 44, MSc (Dal.) 46, PhD (Wisc.) 49
Neish, A.C.,[†] BSc (McGill) 38, MSc 39, PhD 42, FRSC 60, FRS 71
Neufeld, A.H., BSc (Man.) 34, MSc 35, PhD 37, MD (McGill) 50
Newton, R., BSA (McGill) 12, PhD (Minn.) 23, FRSC 30
Nicholson, T.F., MB (Tor.) 26, BSc 28, PhD 34
Nigam, V.N., BSc (Lucknow), MSc 52, PhD (Bombay) 56
Noble, R.L., MD (Tor.) 34, PhD (London) 37, DSc 47, FRSC 50

Odell, A.D., BSc (UNB) 34, PhD (Tor.) 38
O'Neill, A.N.,[†] BA (UBC) 43, MA 45, PhD (Ohio State) 50

Pagé, E., BSA (Montreal) 36, PhD (Cornell) 40, FRSC 55, MBE
Patrick, S.J., BA (Tor.) 42, MA 44, PhD 46
Patterson, A.R.P., BA (UBC) 50, MA 52, PhD 56
Pearce, R.H., BSc (UWO) 46, MSc 48, PhD 51
Perlin, A.S., BSc (McGill) 44, MSc 46, PhD 49, FRSC 69
Pett, L.B., BSA (Tor.) 30, MA 32, PhD 34, MD (Alta.) 42
Polglase, W.J., BA (UBC) 43, MA 44, PhD 48
Pringle, R.B., BSc (Alta.) 44, MSc (McGill) 45, PhD 47
Proulx, P.R., BSc (McGill) 59, MSc 60, PhD 62
Pugsley, L.I., BA (Acadia) 27, MSc (McGill) 30, PhD 32, FRSC 62
Purves, C.B.,[†] BSc (St Andrews) 23, PhD 29, FRSC 49

Quastel, J.H., BSc (London) 21, DSc 26, PhD (Cantab.) 24, FRS 50, FRSC 53

Rabinovitch, I.M., MD (McGill) 17, DSc 32, FRCP(C), OBE
Robertson, T.B.,[†] BSc (Adelaide) 05, PhD (Cal.) 07
Robinson, A.D.,[†] BA (Sask.) 25, MA 27, PhD (Minn.) 30
Rose, D., BSc (Alta.) 39, MSc 41, PhD (Tor.) 46
Rossiter, R.J., BA (West Australia) 34, BA (Oxon.) 38, DPhil 40, BM, BCh 41, FRSC 54

List of biochemists cited 119

Rubin, L.J., BASc (Tor.) 38, MASc 39, PhD 45
Rubinstein, D., BSc (McGill) 49, PhD 53, MD 57
Rudney, H., BA (Tor.) 47, MA 48, PhD (West. Reserve) 52
Ruttan, R.F.,[†] BA (Tor.) 81, MD, CM (McGill) 84, FRSC 95

Sabry, Z.I., BSc (Cairo) 52, MSc (Mass.) 54, PhD (Penn. State) 57
Scholefield, P.G., BSc (Wales) 44, MSc 46, DSc 60, PhD (McGill) 49
Scott, D.A.,[†] BA (Tor.) 20, MA 22, PhD 25, FRSC 39, FRS 49
Sehon, A.H., BSc (Manchester) 48, MSc 50, PhD 51, DSc 65, FRSC 69
Selye, H., MD (Prague) 29, PhD 31, DSc (McGill) 42, FRSC 41
Siminovitch, L., BSc (McGill) 41, PhD 44, FRSC 65
Simms, R.P.S., BSc (McGill) 47, PhD 50
Simpson, F.J., BSc (Alta.) 44, MSc 46, PhD (Wisc.) 52
Simpson, G.E.,[†] BS (Ill.) 13, AM (West. Reserve) 15, PhD (Yale) 20
Sinclair, R.G.,[†] BA (Queen's) 24, PhD (Rochester) 28, FRSC 43
Smillie, L.B., BSc (McMaster) 50, MA (Tor.) 52, PhD 55
Smith, D.B., BA (UBC) 39, MA 41, PhD (Tor.) 50
Smithies, O., BA (Oxon.) 46, MA, DPhil 51
Snell, J.F.,[†] BA (Tor.) 94, PhD (Cornell) 98
Solomon, S., BSc (McGill) 47, MSc 51, PhD 53, FRSC 74
Sourkes, T.L., BSc (McGill) 39, MSc 46, PhD (Cornell) 48, FRSC 71
Speakman, H.B.,[†] BSc (Manchester) 14, MSc 15, DSc 28, FRSC 51
Spencer, E.Y., BSc (Alta.) 36, MSc 38, PhD (Tor.) 41
Spencer, J.H., BSc (St Andrews) 55, PhD (McGill) 60
Spenser, I.D., BSc (Birm.) 48, PhD (London) 52
Stewart, H.B., MD (Tor.) 44, PhD 50
Strickland, K.P., BSc (UWO) 49, MSc 50, PhD 53

Tailleur, P., BSc (Laval) 59, DSc 62
Tarr, H.L.A., BSA (UBC) 26, MSA 28, PhD (McGill) 31, PhD (Cantab.) 34, FRSC 57
Tener, G.M., BA (UBC) 49, MS (Wisc.) 51, PhD 53
Thomson, D.L.,[†] MA (Aberdeen) 21, BSc 24, PhD (Cantab.) 28, FRSC 36
Tisdall, F.F.,[†] MB (Tor.) 16, MD 21, FRCP, FRCP(C)
Trussell, P.C., BSA (UBC) 38, MS (Wisc.) 42, PhD 43
Tuba, J., BSc (Sask.) 32, MSc 37, PhD (Tor.) 44

Vandenheuval, F.A., BSc (Brussels) 32, MSc 34, PhD (London) 38
Venning, Eleanor, BA (McGill) 20, MSc 21, PhD 33, FRSC 55
Verly, W.G., MD (Liège) 47

Vickery, H.B., BSc (Dal.) 15, MSc 18, PhD (Yale) 22
Vining, L.C., BSc (UNZ) 48, MSc 49, PhD (Cantab.) 51, FRSC 74
von Rudloff, E.M., BSc (Pretoria) 48, MSc 50, DSc 53

Wainwright, S.D., BA (Cantab.) 47, PhD (London) 50
Wasteneys, H.,[†] PhD (Columbia) 16, FRSC 30
Watson, E.M.,[†] MD (UWO) 19, MSc 27, FRCP(C) 31, FACP 48
Watson, R.W., BA (Tor.) 25, MA 26, PhD 42
Wetter, R.L., BSc (Alta.) 44, MSc 46, PhD (Wisc.) 50
Whitaker, D.R., BSc (Man.) 41, PhD (London) 48, DSc 63
White, F.D.,[†] ARTC (Glasgow) 13, PhD (Edin.) 23
Williams, G.R., BSc (Liverpool) 49, PhD 51, DSc 69
Wood, A.J., BSA (UBC) 35, MSA 38, PhD (Cornell) 40
Wood, J.D., BSc (Aberdeen) 51, PhD 56
Woodward, J.C., BSA (McGill) 30, MSc (Cornell) 32, PhD 34
Woolley, D.W.,[†] BSc (Alta.) 35, MSc (Wisc.) 36, PhD 38
Wynne, A.M.,[†] BA (Queen's) 13, MA 15, PhD (Tor.) 25, FRSC 43

Yamada, E.W., BSc (UWO) 45, MSc (McGill) 47, PhD (UWO) 51
Young, E.G., BA (McGill) 16, MSc 19, PhD (Cantab.) 21, FRSC 35
Young, L., BSc (Royal Coll. Sci.) 32, PhD (London) 35, DSc 45

Zbarsky, S.H., BA (Sask.) 40, MA (Tor.) 42, PhD 46

Index of names

Ackman, R.G. 61
Adams, G.A. 56, 57, 91
Alcock, A.W. 70
Allardyce, W.J. 37
Allmark, M.G. 66
Anderson, E.M. 17
Anderson, J.A. 71, 91
Archibald, R.M. 10, 100
Armstrong, A.R. 98
Ayre, C.A. 71

Babineau, L.-M. 31, 40, 94
Baer, E. 54, 83, 84, 90, 91
Baigent, M. 96
Bailey, K. 19
Ballantyne, R.M. 96
Banting, F.G. 16, 78–83, 90, 98
Barbeau, A. 93
Baril, G.H. 6, 29–30, 39, 40
Bauman, L. 9
Bayley, S.T. 56
Beall, D. 22
Beare-Rogers, J. 96
Beaton, G.H. 25, 95, 96

Beatty, S.A. 61, 95
Begg, R.W. 17, 27, 34–5
Begin-Heick, N. 94
Bell, J.M. 96
Belleau, B.R. 35
Belzile, R. 96
Benoiton, N.L. 36
Bensley, E.H. 47, 96
Benson, C.C. 24
Berlinguet, L. 31, 40, 94
Bernardi, G. 56
Best, C.H. 16, 24, 52, 78–82, 83, 85, 90
Beveridge, J.M.R. 20, 32–3, 41, 93, 96
Biely, J. 37, 96
Billingsley, L.W. 17
Birchard, F.J. 70
Bishop, C.T. 55, 56, 57, 91
Blanchaer, M.C. 26, 40, 94
Blanchet, R. 93
Bligh, E.G. 61
Bliss, S. 10
Bloor, W.R. 32, 98
Bois, E. 6, 31, 39, 40

Borsook, H. 21, 22
Bouthillier, L.-P. 30, 36, 40, 94
Branion, H.D. 22, 69, 94
Brassard, M. 77
Bridger, W.A. 94
Brisson, G.J. 96
Brocklesby, H.N. 16, 61, 62-3, 67
Brown, A. 69
Brown, M. 32
Browne, J.S.L. 14, 17, 90, 93
Burley, R.W. 56
Butler, G.C. 19, 22, 28, 55, 57, 60, 93, 94, 95
Butler, M.R. 27

Caldwell, B.B. 66
Cameron, A.T. 6, 8-9, 26, 39, 40, 89
Cameron, I. 7
Campbell, J. 22, 85
Campbell, J.A. 65, 66, 96
Campbell, W.R. 20
Cantero, A. 88
Carroll, K.K. 17, 91, 96
Carter, N.M. 16, 62, 63
Casselman, W.G. 85
Chan, A.P. 73
Chapman, R.A. 66
Charles, A.F. 52, 53
Chittenden, R.H. 5
Cinader, B. 87-8
Cipriani, A.J. 59, 60
Clark, W.M. 105
Cohen, S.L. 22
Collier, H.B. 6, 18, 22, 34, 40, 41
Collip, J.B. 6, 11, 12, 14, 16-18, 19, 30, 37, 39, 40, 66, 78-80, 89, 90, 97
Colter, J.S. 18-19, 40, 77
Colvin, J.R. 56
Common, R.H. 49, 96

Connell, G.E. 23, 41, 94
Cook, A.S. 10, 51
Cook, W.H. 55, 56, 94
Copp, D.H. 89
Cornell, B.S. 82
Courtney, A. 85
Craig, B.M. 57
Craine, A. 32
Crampton, E.W. 49, 96
Crocker, B.F. 22

Dale, H. 81, 84
Daoust, R. 88
Darrach, M. 6, 38, 40, 94
Dauphinee, J.A. 20, 21, 41, 45
Davis, G.R. 96
deLamirande, G. 88
Delory, G.E. 26
Demers, J.M. 96
Denstedt, O.F. 11-12, 17, 62, 93, 94
Depocas, F. 56, 60
Despointes, R.H. 6, 41
D'Iorio, A. 35, 41, 90, 94, 97
Dixon, G.H. 38, 90
Downs, A.W. 93
Drake, T.G. 10, 69, 86
Dvornik, D.M. 51
Dyer, W.J. 61

Eadie, G.S. 20
Eagles, B.A. 10, 37
Ebbs, J.H. 69
Edwards, O.E. 89
Elliott, K.A.C. 12-13, 40, 77, 78, 90, 97
Emslie, A.R.G. 22, 72
Engel, C. 17
Ettinger, G.H. 93
Ettori, J. 35, 41
Evans, E.V. 96

Evans, H.S. 64, 66
Evelyn, K.A. 15, 105

Farmilo, C. 66
Feltham, L.A.W. 6, 40
Ferguson, J.K. 93
Finn, D.B. 61, 62
Fischer, H.O.L. 54, 83-4, 91
Fisher, A.M. 52, 53
Fitzgerald, J.G. 52
Flavelle, J. 85
Fleishmann, G. 73
Flexner, A. 5
Folin, O. 26, 106
Fowler, A.F. 47
Franklin, M. 77
Fraser, M.J. 13-14, 77, 97
Freudenberg, K.J. 58
Fritz, C.W. 64
Fritz, I.B. 85

Gaebler, O.H. 10
Gairns, S. 82
Gaudin, C. 31
Gaudry, R. 31, 51, 93, 95
Gavin, G. 25
Geddes, W.F. 71, 91
Gianetto, R. 30
Gibbons, N.E. 57
Gillespie, R.J. 85
Gilmore, C.R. 8
Gingras, J.R. 6, 31, 40
Girdwood, G.P. 9
Givens, M.H. 20
Godin, C. 31, 94
Gorham, P.R. 56
Gornall, A.G. 20, 21, 41, 45-6
Gosselin, G. 30
Gowe, R.S. 71
Grace, N.H. 57, 74

Graham, A.F. 13, 40
Graham, W.R. 22
Grant, G.A. 10, 27, 51
Greenway, H.F. 67
Griffiths, J.C. 41, 47
Guest, G.H. 48
Gurd, F.R.N. 100

Hagen, P.B. 26, 33, 40, 41
Hall, R.H. 6, 39, 40, 94
Hanes, C.S. 23, 38, 41
Harding, V.J. 9, 10, 18, 27, 37, 39,
    41, 45, 84, 86, 100
Harington, C.R. 53, 60, 84
Harlow, C.M. 17
Harris, N.M. 65
Harrison, W.A. 89
Hart, J.S. 56
Haskins, R.H. 57
Hawkins, W.W. 27, 59, 96
Hayes, F.R. 29
Heard, R.D.H. 11, 12, 15, 22, 90
Helleiner, C.W. 22, 28, 40
Henderson, J.F. 87
Henderson, V.E. 82, 93
Hibbert, H. 16, 58, 62, 63
Hingerty, D. 35, 40
Hlynka, I. 71
Hochster, R.M. 72, 73, 77, 94
Holmes, H.L. 89
Hopkins, F.G. 21, 106
Hopper, W.C. 67
Hosein, E.A. 13, 90
Howland, J. 69
Hunter, A. 20-1, 28, 34, 41, 45, 48,
    52, 89
Hunter, G. 18, 19, 37, 40, 67, 84
Hurst, R.O. 22, 33

Idler, D.R. 61, 90

124 Index of names

Inman, W.R. 27
Irvine, G.N. 71
Irving, L. 23
Israels, L.G. 86

Jackson, S.H. 10, 69, 85
Jean, M. 31, 40
Jellinck, P.H. 33, 41
Johns, H.E. 87
Johnston, W.W. 22
Jones, J.K.N. 33, 91
Joubert, F.J. 56
Jukes, T.H. 22, 99–100

Kamen, M.D. 100
Kaneda, T. 75
Kates, M. 36, 57, 97
Kay, C.M. 19
Kay, H.D. 22, 100
Keirstead, K.F. 31
Khorana, H.G. 38, 75
King, E.J. 26, 84, 98–9
Kirkwood, S. 39, 100
Krotkov, G. 91
Kuksis, A. 85
Kulka, M. 89

Labarre, J. 30
Labrie, F. 31
Lachance, J.P. 30
Laidler, K.J. 36
Lamy, F. 41
Lancaster, H.M. 65, 66
Lang, J.M. 85
Larmour, R.K. 57
Lathe, G.H. 100
Laughland, D.H. 22, 94, 95
Layne, D.S. 35, 41
Leathes, J.B. 6, 41, 44–5
Leblond, C.P. 89

Ledingham, G.A. 57
Lemieux, R.U. 57, 91
Lentz, C.P. 57
Lloyd, L.E. 49, 96
Logan, J.F. 32
Long, C.N. 14
Lothrop, A.P. 32, 39
Lozinski, E. 51–2
Lucas, C.C. 20, 28, 32, 82, 84, 85
Lucas, G.H. 82
Luck, J.M. 99
Lupien, P.J. 96
Lusk, G. 106

Macallum, A.B. Jr. 26, 41
Macallum, A.B. Sr. 5–7, 10, 11, 16, 17, 20, 24, 39, 40, 41, 55
Macfarlane, T. 64, 66
MacKenzie, J.J. 7
MacLean, D.B. 89
Macleod, J.J. 16, 78–80
MacPherson, C.F.C. 27, 77
Macpherson, L.B. 28–9
Macpherson, M. 59
MacRae, H.F. 49
McArthur, C.S. 20, 34, 41
McCalla, A.G. 49, 92
McCarter, J.A. 22, 28, 40, 60, 87, 94
McConnell, W.B. 58
McDonald, B.E. 96
McFarlane, W.D. 20, 48–9, 66, 67
McGill, A. 64–5, 66
McHenry, E.W. 24–5, 67, 93, 94, 95, 96
McLachlan, J.L. 59
McLean, W.F. 53
McMurrich, J. 7
McPhail, M.K. 17
McPhedran, A. 7
Madsen, N.B. 19, 94
Manery (Fisher), J. 23

Manning, R.J. 34, 39, 93
Manske, R.H. 89
Marion, L. 89
Marko, A.M. 22
Marmur, J. 100
Marrian, G.F. 22, 90
Martel, F. 31
Martin, H. 73
Martin, W.G. 56
Mathews, A.P. 7, 107
Mawson, C.A. 60
Meakins, J.C. 14
Mendel, L.B. 107
Menon, K.K. 85
Meredith, W.O. 71
Middleton, E.J. 66, 96
Migicovsky, B.B. 70
Miller, F.R. 78
Moloney, P.J. 52
Monagle, J.E. 68
Mongeau, E. 96
Mookerjea, S.S. 85
Moorehouse, V.H. 93
Morais, R. 88
Morgan, J.F. 86, 87, 97
Morrell, C.A. 22, 65, 66
Morrell, J.A. 20
Morrison, A.B. 66, 96
Moshier, H.H. 19
Mounce, I. 8
Mulock, W. 82
Murnaghan, M. 35, 40
Murray, T.K. 65
Murthy, M.R.V. 32

Nadeau, G. 31
Nadeau, H. 67
Neelin, J.M. 22, 94
Neilands, J.B. 27, 100
Neish, A.C. 16, 57, 58

Neufeld, A.H. 17, 41, 46
Newton, R. 49, 55, 67, 71
Nicholson, T.F. 10, 46
Nigam, V.N. 88
Noble, R.L. 17

Odell, A.D. 22
O'Neill, A.N. 59
Orr, J.B. 72
Orr, M.D. 52
Osler, W. 5

Pagé, E. 30, 95
Page, I.H. 77, 78
Paterson, J.C. 17
Patrick, S.J. 28
Patterson, A.R.P. 87
Patterson, J.M. 67, 85
Pearce, R.H. 47
Perlin, A.S. 91
Peters, R.A. 21
Pett, L.B. 22, 67, 68, 96
Phillips, W.E. 96
Poirier, L. 88
Polglase, W.J. 38, 40
Pringle, R.B. 73
Proulx, P.R. 36
Pugsley, L.I. 17, 66
Purves, C.B. 57, 91

Quastel, D.M. 77
Quastel, J.H. 11, 73, 76–7, 78, 86, 94

Rabinovitch, I.M. 47, 76
Rhodes, A.J. 86
Rice, F.A. 27
Ridout, J.H. 85
Robertson, E.C. 69, 85
Robertson, T.B. 20, 21, 41
Robinson, A.D. 50

Index of names

Rose, B. 15
Rose, D. 57
Rosenfield, B. 85
Rossiter, R.J. 27, 41, 90, 94, 95
Rubin, L.J. 54
Rubinstein, D. 13
Rudney, H. 100
Ruttan, R.F. 5, 9-10, 27, 39, 64

Sabry, Z.I. 68-9
Sahasrabudhe, M. 72
Sallans, H.R. 71
Salter, J.M. 85
Sauer, F. 71
Scholefield, P.G. 76, 77, 94
Scott, D.A. 20, 52-3, 90
Scott, J.W. 40
Sehon, A.H. 15-16
Selye, H. 17, 30-1
Sheinin, V.P. 94
Shutt, F.B. 69
Siminovitch, L. 60, 87
Simms, R.P.S. 72
Simpson, F.J. 57, 59
Simpson, G.E. 10
Sinclair, R.G. 6, 32, 41
Smillie, L.B. 19, 22, 94
Smith, D. 24
Smith, D.B. 22, 56
Smith, D.G. 59
Smithies, O. 52
Snell, J.F. 48
Solomon, S. 15
Sourkes, T.L. 77, 90
Speakman, H.B. 22
Spencer, E.Y. 73
Spencer, J.H. 94
Spenser, I.D. 39
Stevenson, J.A. 17
Stewart, A. 67

Stewart, H.B. 27, 41
Strickland, K.P. 27, 41
Swaine, J.M. 67
Sylvestre, J.E. 67

Tailleur, P. 31, 40
Tait, J. 93
Tarr, H.L.A. 16, 63
Taylor, A. 59
Taylor, B. 17
Taylor, N.B. 82
Tener, G.M. 38
Terroine, E.F. 21
Thatcher, F.S. 66
Thompson, R.H. 27, 99
Thomson, D.L. 11, 12, 40
Thorvaldson, T. 74
Tisdall, F.F. 67, 69, 85, 86
Toby, C.G. 17
Tomlinson, N. 63
Trussell, P.C. 75
Tuba, J. 19

Valenta, Z. 89
Vandenheuval, F.A. 72
Van Slyke, D.D. 21
Venning, E. 14-15, 90
Verly, W.G. 30, 40
Vickery, H.B. 99
Vining, L.C. 29, 57, 58, 59
Vlassopoulos, V. 6, 35, 40
von Rudloff, E.M. 58

Wainwright, S.D. 28
Walker, W. 9
Wasteneys, H. 21-2, 41, 48, 67
Watson, D.W. 27
Watson, E.M. 6, 41, 46
Watson, R.W. 56
Webber, R.V. 27

Webley, D.M. 77
Wetter, R.L. 58
Whitaker, D.R. 36, 56, 94, 95
White, F.D. 26, 40
Wiberg, G.S. 66
Wiesner, K. 89
Williams, G.R. 23, 41, 94
Wood, J.D. 34, 41
Woods, D.D. 28
Woodward, J.C. 70
Woolley, D.W. 99
Wright, R. 6

Wynne, A.M. 22, 23, 41, 67, 87, 94

Yamada, E.W. 26
Yaphe, W. 59
Yip, C.C. 85
Young, E.G. 6, 10, 26, 27, 28, 38, 39, 40, 41, 58, 59, 67, 93, 94, 95
Young, L. 22, 28
Young, R.J. 85

Zbarsky, S.H. 22, 38, 60, 94

Index of institutions

Acadia University 33, 42, 43
Agricultural Research Institutes 69-74
Alberta, University of 6, 16-20, 39, 40, 41, 42, 43, 44, 49, 67, 71, 86, 87, 91, 92
Allan Memorial Institute of Psychiatry 77
Animal Research Institute 71-2
Atlantic Regional Laboratory 58-9
Atomic Energy of Canada 59-60
Ayerst Laboratories 51

Banting and Best Department of Medical Research 81, 82-4, 85, 91
Banting Institute 78-82, 85
Banting Research Foundation 82, 101
Best Institute 85
Bishops University 42
Brandon College 42, 43, 84
British Columbia, University of 6, 36-8, 40, 42, 43, 90
British Post-Graduate Medical School 26, 99
Brock University 42

Calgary, University of 42
Cambridge University 44
Canada Packers 53-4
Canadian Biochemical Society 93-5
Canadian Physiological Society 93
Carleton University 43
Chemistry and Biology Research Institute 72-3
Collip Medical Research Laboratory 17, 27
Concordia University 42
Connaught Laboratories 52, 53, 80, 83

Dalhousie University 6, 26, 27-9, 40, 41, 42, 43, 44
Defence Research Board 34, 101
Department of Agriculture 69-73, 92
Department of the Environment 61
Department of National Health and Welfare 64, 101

Fisheries Research Board 60-3
Food and Drug Laboratories 64-6
Food Research Institute 72

Index of institutions 129

Forest Products Laboratories 63–4
Freshwater Institute 60, 61
Frosst, C.E. & Co. 51

Grain Research Laboratory 70–1, 91
Guelph University 42

Hospital for Sick Children Research
  Institute 85–6

Laval University 6, 29, 30, 31–2, 40,
  42, 43
Loyola University 43

Macdonald College 48, 49
McGill-Montreal General Hospital
  Research Institute 76–7, 86, 90
McGill University 5, 6, 9–16, 17, 18,
  19, 40, 42, 43, 47, 89, 90
McMaster University 6, 39, 40, 42, 43
Manitoba Cancer Research Unit 13, 86
Manitoba, University of 6, 8, 26, 40,
  42, 43, 44, 46, 50
Medical Research Council 102, 103
Memorial University of Newfoundland
  6, 40, 42
Montreal Cancer Institute 86, 88
Montreal Neurological Institute 12,
  78, 90
Montreal, University of 6, 29–31, 40, 43
Mount Allison University 43

National Cancer Institute 86–8, 101,
  102
National Research Council 55–9, 102,
  103
New Brunswick, University of 43
Nova Scotia Agricultural College 49, 58
Nutrition, Department of 67–9

Nutrition Society of Canada 95–7

Ontario Agricultural College 43, 48
Ontario Cancer Institute 13, 28, 86, 87
Ontario Research Foundation 22, 23
Ottawa, University of 6, 35–6, 40, 41,
  42, 43, 44

Pacific Fisheries Experimental Station
  12, 61–3
Plant Research Institute 73
Prairie Regional Laboratory 57–8, 59
Provincial Research Laboratories 74–5

Quebec, University of 31
Queen's University 6, 32–3, 34, 35,
  41, 42, 43, 44, 91, 93, 98

Royal Victoria Hospital 13, 14, 15

St Francis Xavier University 42, 43
Saskatchewan, University of 6, 34–5,
  41, 42, 43, 86
Sherbrooke, University of 6, 41, 42
Simon Fraser University 42, 43, 44
Sir George Williams University 43

Toronto, University of 5, 6, 20–5, 26,
  41, 42, 43, 44–6, 90

Victoria, University of 6, 42

Waterloo University 42, 43, 44
Western Ontario, University of 6, 26–7,
  28, 41, 42, 43, 44, 46–7, 90
Windsor, University of 42

York University 42, 43

www.ingramcontent.com/pod-product-compliance
Lightning Source LLC
Chambersburg PA
CBHW060452080526
44584CB00015B/1414